Rewriting the History
in North America 1607–1861

Nerida Ellerton · M.A. (Ken) Clements

Rewriting the History of School Mathematics in North America 1607–1861

The Central Role of Cyphering Books

Foreword by Jeremy Kilpatrick

Springer

Dr. Nerida Ellerton
Illinois State University
Campus Box 4520
Normal, IL 61790-4520
USA
ellerton@ilstu.edu
nerida@ellerton.net

Dr. M.A. (Ken) Clements
Department of Mathematics
Illinois State University
Normal, IL
USA
clements@ilstu.edu

ISBN 978-94-017-8095-7   ISBN 978-94-007-2639-0 (eBook)
DOI 10.1007/978-94-007-2639-0
Springer Dordrecht Heidelberg London New York

© Springer Science+Business Media B.V. 2012
Softcover reprint of the hardcover 1st edition 2012
No part of this work may be reproduced, stored in a retrieval system, or transmitted in any form or by any means, electronic, mechanical, photocopying, microfilming, recording or otherwise, without written permission from the Publisher, with the exception of any material supplied specifically for the purpose of being entered and executed on a computer system, for exclusive use by the purchaser of the work.

Printed on acid-free paper

Springer is part of Springer Science+Business Media (www.springer.com)

Foreword

In an encyclopedia by the German monk Gregor Reisch first published in 1503, a well-known woodcut *Typus Arithmeticae* portrays Dame Arithmetic presiding over what is apparently a competition between Boethius, who is writing calculations in Hindu-Arabic numerals using a pen, and Pythagoras, who is using a counting board. Arithmetic gazes approvingly toward the confident algorist Boethius, who seems to have finished his calculation, while the abacist Pythagoras looks stumped. Although Reisch has made a curious choice of competitors, he has correctly anticipated the outcome of the contest. The replacement of the abacus by written algorithms for computing in Western Europe took several centuries and was not completed until the late sixteenth century. Once writing had been established, however, as the way arithmetic—not to mention the rest of mathematics—was to be done, it enjoyed a long reign and spread quickly to the colonies.

North American historians of education have not always attended to the prominent role that writing played in teaching pupils mathematics. Instead, they have focused attention on the heavy reliance that American mathematics teachers have had on textbooks. Certainly that reliance has been profound. After the U.S. Civil War, the rise of the age-graded school coupled with the rapid Western expansion of education and remarkable growth of public high schools led to a shortage of well-educated teachers, especially teachers of mathematics. Commercially published mathematics textbooks, along with so-called ancillary materials, were written and marketed to help semi-skilled teachers cope with mathematical ideas that they might not understand very well. Although it is impossible to determine whether, over the past century and a half, mathematics teachers in North America have relied more heavily on textbooks than teachers in other countries, their textbooks have certainly been bulkier, more elaborate, and more heavily scripted than textbooks elsewhere. Moreover, textbooks have been the principal conduits for attempts to change the mathematics curriculum, as shown by federally funded projects to create new textbooks during the roughly two decades of the new math era and the more recent so-called standards era.

A preoccupation with textbooks has led historians to ignore or devalue the way cyphering books were used to learn mathematics in North America during the two and a half centuries preceding 1861. They have tended to assume that in a time before textbooks became commonplace in mathematics classrooms, pupils simply copied from each other or from whatever book their teacher had without thinking too much about what they were doing. The pedagogical value of writing as a means of formulating and consolidating one's mathematical thinking is easily overlooked once printed materials become readily available. But for centuries, the principal way that young scholars learned and remembered mathematics was by writing it carefully in their own treasured cyphering books.

As Ellerton and Clements show, the cyphering approach to learning mathematics has both an extensive history and a complex rationale. In the approach, the learner was to become an independent problem solver by learning to identify problems of various types, learning the rules for their solution, solving such problems, having each solution checked by a tutor or teacher, and only when the solution was correct, copying that solution into a cyphering book using exemplary calligraphy. The handwriting process was not meant to be only copying; it was to be an accompaniment to thinking. The learner was not simply inscribing the solution on to paper; he or she was inscribing it into memory.

Cyphering books are a much neglected resource for understanding and appreciating the early history of North American mathematics education. The present book marks a major advance in putting that resource to effective use as well as in raising some challenging questions for subsequent research.

Athens, Georgia   Jeremy Kilpatrick

Abstract

An analysis is presented of the historical, societal, and mathematics education practices that defined what we call the "cyphering tradition" which prevailed in the North American (mainly British) colonies between 1607 and 1776, and then in the United States of America for most of the period between 1776 and 1861. The main data set comprises 270 handwritten cyphering books, of which 212 originated in North America between 1701 and 1861.

The following five research questions provided the foci for the study:

1. Where and when were the cyphering books prepared, and by whom?
2. Were the manuscripts consistent with *abbaco* traditions, especially in relation to mathematical content and sequencing, genre, and "writing as also arithmeticke"?
3. What theoretical base, if any, can be identified to encapsulate the educational purposes of the cyphering books?
4. Why, between 1840 and 1861, was there such a sharp decline in the use of the cyphering approach in the Unites States?
5. What were the main educational advantages and disadvantages of the cyphering approach?

It was found that most of the North American manuscripts were prepared between 1780 and 1850 in the North-Eastern regions of what is now the United States of America. About 83 percent of them were prepared by males, usually during winter months. After 1840, the extent of the use of cyphering books declined sharply. Three reasons are identified for the decline: first, cyphering books were no longer regarded as important in relation to evaluating the quality of an instructor's teaching or of a student's learning; second, key state education leaders favoured pedagogical reform by which mathematics would be taught to graded classes, rather than to individuals; and third, young teachers graduating from state normal schools discouraged the use of the cyphering approach, especially in the new high schools that were being established.

Preface

The term "rewriting" in the title of this book merits comment. In this work we argue that published histories dealing with the teaching and learning of school mathematics in the North American colonies, and during the first 85 years of the United States of America, have not recognized the fundamental importance of what we call the "cyphering tradition," and have over-emphasized the role of textbooks. We provide evidence from numerous primary sources that printed texts were less used and, from a mathematics teaching and learning perspective, were less important, than handwritten manuscripts prepared by learners. These handwritten documents were referred to as "cyphering books" by the students who prepared them, and by their instructors. We argue that a cyphering tradition, imported from European nations to the North American colonies, provided both the practical and the theoretical base for early North American mathematics education.

After reviewing the European background to the cyphering tradition, we show how the cyphering tradition was translated into the North American colonies, where it remained the dominant influence on mathematics teaching and learning until well into the 19th century.

Historians of mathematics education in North America have often written more from the college than from the school perspective. Furthermore, because it has been relatively easy to access a wide range of mathematics textbooks, and descriptions of mathematics courses used in various colleges in the 17th, 18th centuries and early 19th centuries, histories have often been more concerned with elaborating the *intended* rather than *implemented* or *attained* curricula.

The principal data set, on which much of the analyses summarized in this book were based, is the largest-known single collection of North American cyphering books. We have linked our analyses of these data with analyses of other published and unpublished materials, including textbooks. Most of the 212 cyphering books in the principal data set were prepared by school students and not college students, and hence we have been able to look closely at what went on in the name of mathematics for students who did not necessarily intend to proceed to college.

We wish to thank librarians and archivists in the Phillips Library (Peabody Essex Museum, Salem, Massachusetts), the Butler Library (Columbia University, New York), the Clements Library (University of Michigan), the Houghton Library

(Harvard University), the American University (Washington, DC), the Wilson Library (University of North Carolina at Chapel Hill), the Special Collections Research Center in the Swem Library at the College of William and Mary, and the Rockefeller Library (both in Williamsburg, Virginia), the New York Public Library, Guildhall Library (London), the London Metropolitan Archives, and the Milner Library (Illinois State University) for locating relevant manuscripts, artifacts, and books for us.

We would particularly like to thank Dr George Seelinger, the Head of the Mathematics Department at Illinois State University (in which we both work) for encouraging us in our research endeavors. We also thank Dr Roger Eggleton for insightful comments which helped us finalize the title of the book, and improve the wording of parts of this preface.

Normal, Illinois

Nerida Ellerton
M.A. (Ken) Clements

Contents

1 The Historical Challenge 1
   Clarifying Concepts, and Associated Nomenclature 2
      Schools 2
      Cyphering Books 3
      Abbacus School 4
      Mathematical Activity and Mathematics Education 5
      Nomenclature 5

2 Development of a Tradition 7
   Origins of the Cyphering Approach to Mathematics Education
   in Western Europe 7
   The *Abbacus* Manuscript Tradition in Western Europe 1200–1600 ... 9
      The Emergence of the Italian *Abbaco* Tradition 9
   Van Egmond on the *Abbaci* and the History of Mathematics
   Education 11
      The Content of the *Abbaci* 14
   Mathematics Beyond Reckoning 16
   Changing Attitudes Toward the Value of Mathematical Studies 17
   Linking Writing Education with Mathematics Education 19
      The Cyphering Approach 19
   The Role of Mathematics in Education 23
   The Royal Mathematical School at Christ's Hospital, London 25
      "Writing as also Arithmeticke" at Christ's Hospital 29
      The Christ's Hospital Model for Mathematics Education 30
   Toward a Theoretical Model for the Cyphering Approach 31
      A Structure-Based, Problem-Solving Model Implicit
      in the Cyphering Approach 32
      A Case in Point: Teaching and Learning the Various Forms
      of the "Rule of Three" 33
      Teachers and Students Responding to Difficult Teaching
      and Learning Environments 34

3	**Translating the Cyphering-Book Tradition to North America**	37
	Educational Realities in the New World	37
	Learning to Count and Measure in the Early European Settlements of North America	42
	Why Study Arithmetic Beyond the Four Operations on Whole Numbers?	46
	The Practical Emphases in Colonial Cyphering Books	47
	Practical Emphases in Arithmetic and Navigation Cyphering Books . .	60
	Practical Emphases in the Phillips Library Cyphering Books	61
	Ethnomathematical and Methodological Considerations	70
	Comments on the Effectiveness of the Cyphering Approach	73
	The Evaluative Function of Cyphering Books	76
4	**Formulating the Research Questions**	79
	Our Journey	79
	The Research Questions That Guided the Study	84
5	**The Principal Data Set**	87
	The Extent of the Collection	87
	Analyses of the Principal Data Set	88
	Details Relating to the Cyphering Books	88
	The Three Periods	92
	The Gender of Writers	93
	Locations of Manuscripts	95
	The Mathematical Content in the Cyphering Books	96
	Algebra, Geometry, Trigonometry, Surveying and Navigation	100
	Genre Considerations	102
	Relationships Between Mathematics Textbooks and Cyphering Books	107
	The Influence of Commercially-Printed Texts on the Preparation of Cyphering Books	107
	In Defense of the Cyphering Approach to Mathematics Education	113
	The Nature and Role of Recitation	116
6	**Ending a 600-Year Tradition: The Demise of the Cyphering Approach**	119
	Warren Colburn and the Challenge to the Cyphering Tradition	119
	Monitorial Systems for School Mathematics	121
	The Continuing Emphasis on Cyphering 1820–1840	121
	The Demise of the Cyphering Approach	124
	Why Did the Cyphering-Book Tradition Lose Favor During the Period 1840–1865?	125
	Normal Schools, Public High Schools, New Textbook Series, and Cyphering Books	136

7 Conclusions, and Some Final Comments	139
Answers to Research Questions	139
Research Question 1	139
Research Question 2	140
Research Question 3	142
Research Question 4	145
Research Question 5	146
Final Evaluative Comments	148
Recommendations for Further Research	151
Author Biographies	153
Appendix A: Summary of the Cyphering Books in the Principal Data Set	155
Summary of Cyphering Books in the Principal Data Set, North American Colonies/States	157
Appendix B: Cyphering Books Held by the Huguenot Historical Society	185
Item List	186
Item Descriptions	186
Appendix C: Conversion Tables for US Systems of Measurement, Around 1800	191
References	195
Author Index	213
Subject Index	219

List of Figures

Figure 2.1	A page from the *abbacus* manuscript by Andréas Reinhard, 1599	13
Figure 2.2	The title page of William Butler's (1819), *An Introduction to Arithmetic: Designed for the Use of Young Ladies*	22
Figure 2.3	A page from Charles Page's (1825) cyphering book, prepared at Christ's Hospital	27
Figure 3.1	A compound fellowship problem from an early North American cyphering book	50
Figure 3.2	A "reduction" task from, perhaps, a 17th-century North American manuscript	51
Figure 3.3	An example of PCA genre from Thomas Prust's (1702) cyphering book	53
Figure 3.4	Solution to an alligation word problem by David Townsend (1770–1774)	55
Figure 3.5	Tare and trett—in Thomas Burlingame Junior's (1779–1784) cyphering book	56
Figure 3.6	Solution to a "double position" problem—Seth Green Torrey's (1788) manuscript	58
Figure 3.7	Thomas Rowlandson's (1789) depiction of "A Merchant's Office."	72
Figure 5.1	Number of cyphering books per decade (1780–1789 through 1870–1879) in the PDS	90
Figure 5.2	Ratio of cyphering books prepared to millions in the US population per decade, 1780–1789 through 1870–1879, in the PDS	90
Figure 5.3	A page from Sally Halsey's (1767) cyphering book (Appendix A, m/s #4)	94
Figure 5.4	A page from the 1824 cyphering book of Anderson Newman, of Shenandoah, Virginia	99
Figure 5.5	A page from William Turnbull's (1780) (Appendix A, m/s # 13) book	103

Figure 5.6	Statement and solution of two type problems (for subtraction), from Thomas Priest's ("Prust's") (1702) cyphering book (Appendix A, m/s #2)	104
Figure 5.7	A page from the cyphering book prepared by an anonymous student around 1815 (Appendix A, m/s #79) . . .	105
Figure 6.1	Page 32 from Adams (1822), with "blanks" completed by an owner	128
Figure 7.1	Educational rationale for the cyphering approach (Ellerton & Clements, 2009b)	143

List of Tables

Table 3.1	Comparative Table of College Mathematic Entrance Requirements, 1786–1800	78
Table 5.1	Summary of the 212 North American Cyphering Books in the Principal Data Set	89
Table 6.1	Questions on the First U.S. Externally-Set Arithmetic Paper Administered to 308 Public School Students, with Numbers and Percentages Giving Correct Answers (Caldwell & Courtis, 1925)	130

Chapter 1
The Historical Challenge

Abstract Between 1607 and 1840 school mathematics in North America was very different from school mathematics today. Throughout that period, the cyphering tradition dictated the mathematics curriculum and the ways mathematics was taught and learned. Early North American teachers did not stand at the front of the room and teach, and most students, even those studying mathematics, did not own a mathematics textbook. Written examinations of any kind were not used. Most teachers of any branch of mathematics did not have formal qualifications in mathematics. Those studying mathematics copied, into their own cyphering books, rules, cases and examples from existing handwritten cyphering books. Definitions of key terms (e.g., "cyphering book") are given.

Mathematics teachers are often accused of being full of inertia, in the sense that they are reluctant to change curricula or teaching approaches (Taylor, 2002). The standard explanation for this perceived state of affairs is that because teachers' daily professional lives are prescribed by regulations, timetables, entitlements and expectations, they become comfortable working in ways which, they believe, work for them (Lortie, 1975). Furthermore, evaluators tend to assess curricula and teaching approaches in terms of what has occurred in the past, rather than what might be possible if new opportunities (arising out of, say, technological advances) were grasped (Papert, 1980).

Thus, so the argument goes, no matter how much society might think large-scale changes are needed in mathematics teaching and learning, these are very difficult to achieve. From this line of argument, it is not difficult to reach the conclusion that present mathematics teaching methods are likely to be similar to those used 100 or even 200 years ago.

But, in fact, such a conclusion would be well wide of the mark, for mathematics education practices in the United States of America are very different now from what they were in the 18th and early 19th centuries in the colonies and states. In the early 1800s many school-age children in the United States rarely attended school. Of those who did attend, many (especially boys in the New England colonies or states) attended during winter months only, and did not study any mathematics beyond elementary arithmetic. As late as the 1870s, the hero of Mark Twain's *Adventures of Huckleberry Finn* went to school only "three or four months" during which time he learnt the rudiments of reading, writing, spelling, and "the multiplication table up to six times seven is thirty-five" (Clemens, 1885, p. 18).

Two centuries ago, American teachers did not stand at the front of the room and teach, and most students, even those studying mathematics, did not own a mathematics textbook. Written examinations of any kind were not used. Most teachers of any branch of mathematics did not have formal qualifications in mathematics.

In this book we review changes to the principal method of teaching and learning mathematics in North America in the 17th and 18th centuries. Our analysis for the period 1840–1860 reveals that in just two decades a cyphering tradition, that had spread to Europe and America after originating in India and in Arabic nations, and had endured for over 600 years, was done away with in North American schools. We shall offer an explanation for how and why this occurred.

In providing details of the early traditions relating to the teaching and learning of mathematics in the North American colonies and, between 1776 and 1861, within the United States of America, we shall present an account of developments that differs from those offered in other histories. For example, unlike other analyses of school mathematics in colonial times, we will not emphasize the role of mathematics textbooks, arguing that until about 1800 such textbooks were little used in the North American colonies, except in colleges like Harvard and Yale. Our thesis will be that the main teaching and learning aid for mathematics education was the cyphering book, and that instruction and learning were consistent with what we have called the "cyphering tradition." We shall also draw attention to ethnomathematical contexts in which children and adults learned mathematics outside of formal education institutions.

Much of the focus of this book will therefore be on the cyphering tradition as it operated in North America before the 1860s. We shall begin by defining the terms "cyphering book," "*abbacus* school," and "mathematical activity," and then move on to a description of what constituted the cyphering tradition. We shall argue that this tradition was transplanted from Europe to the North American colonies, and had a pronounced influence on mathematics education practices during the first 250 years of European settlement in North America.

Following a historical overview of factors influencing the cyphering tradition, the set of questions that guided the research will be presented, and the principal data set will be summarized. Subsequent analyses will form the basis of our reconstruction of the early history of North American school mathematics.

Clarifying Concepts, and Associated Nomenclature

Schools

The concept of "school" in this book will embrace a wide range of educational settings. Included will be "dame schools," "common schools," "public schools," "subscription schools," "private schools," "evening schools," "writing schools," "high schools," and "apprenticeship schools." Although, sometimes, words like "academy" will be used instead of "school," whenever we use the word "school" (or "academy," etc.) we will be referring to education environments, both formal and informal, in which at least one "teacher" regularly met with a gathering of "students," at an agreed place, for the purpose of helping those students to learn the elementary facts, concepts, and skills associated with at least one particular area of knowledge. Our use of "school" embraces far more than just "private" and "public schools."

Writing in 1921, Robert Seybolt pointed out that evening schools in urban centers had played important roles in the history of education in 17th-, 18th- and early 19th-century North America. Seybolt (1921) wrote of evening schools in colonial New York:

> The available records indicate that the earliest ... offered instruction only in the rudiments—reading, writing and cyphering. It is probable that these were attended exclusively by apprentices. In some instances, adults may have received such evening instruction, but on this matter the records are silent. (p. 630)

Right through the colonial period, schools were an important part of society. Thus, for example, although the population of the "city" of New York around 1700 was only about 5000 (Oswald, 1917), there were several evening schools within the city at that time (Seybolt, 1921).

Note that in this book higher-level colleges—such as Columbia, Harvard, William and Mary, and Yale—will not be regarded as "schools," for in the 17th and 18th centuries, such higher-level institutions were usually sharply distinguished from schools.

Cyphering Books

In this book we argue that before 1840, students in North America who attempted to learn any branch of mathematics usually did so through a cyphering approach (Meriwether, 1907). That statement raises at least two questions: (a) What were the main elements of the cyphering approach? And, (b) Why was this approach used so extensively in the United States until about 1840? We shall answer both of those questions.

We begin by defining a *cyphering book* as a handwritten manuscript with the following four properties:

1. Either the contents were written by a student who, through the act of preparing it, was expected to learn and be able to apply whatever content was under consideration; or, the book was prepared by a teacher who wished to use it as a model that could be followed by students preparing their own cyphering books.
2. All entries in the book appeared in ink—either as handwriting or as illustrations. Headings and sub-headings were presented in a decorative, calligraphic style.
3. The book was dedicated to setting out rules and cases associated with a sequence of mathematical topics, with each topic being followed by problems linked with that topic. The problems were in arithmetic, especially business arithmetic, or

in algebra, or geometry, or trigonometry, or were applications of mathematics in the fields of navigation, surveying, military strategy, etc.
4. The topics covered were sequenced so that they became progressively more difficult. The content also reflected the expectation that, normally, no child less than 10 years of age would be assigned the task of preparing a cyphering book.

Most of the cyphering books in the principal data set (hereafter "PDS") for this study dealt with just one branch of mathematics (e.g., usually arithmetic, but sometimes algebra or geometry, etc.). Occasionally, a cyphering book had entries from several areas of mathematics. In the 18th and 19th centuries in North America, cyphering books were usually made up of unlined, rectangular folio-sized paper sheets (in the 18th century, "rag" paper was often used) with dimensions about 12.5″ by 8″. The pages, grouped as sections, were routinely sewn together to form books (in the form of handwritten manuscripts). Sometimes, protective covers were added. Typically, the first page of a book was beautifully decorated, and indicated the name of the owner as well as the year(s) and location in which the book was prepared.

Most cyphering books dealt with a number of topics. Although, usually, the treatment of any one topic occurred on successive pages, it was not uncommon for topics to be revisited at various times in the same cyphering book. Cyphering books were intended to serve as reference books for future use by those who prepared them.

In addition to our collection of cyphering books we own other handwritten manuscripts that, although intended for school mathematics purposes, do not fit our definition of a cyphering book. Thus, for example, we have an original handwritten manuscript written by an 18th-century author who had prepared it for possible publication as a textbook on geometry, trigonometry, and navigation. We also have a manuscript, prepared by a school administrator during the period 1828–1852, that set out how arithmetic was to be taught in a Canadian school that adopted the Lancastrian monitorial approach. In addition, we have a collection of handwritten documents in which writers concentrated solely on stating and solving mathematics problems—in these manuscripts, introductory summaries, and rules and cases were not stated, and therefore we have not regarded them as cyphering books.

Abbacus School

According to Jens Høyrup (2008), *abbacus* schools, which operated in Italy from the 13th century, were

> … primarily frequented by merchant and artisan youth for … two years (around the age of 11), who were taught the mathematics needed for commercial life: calculation with the Hindu-Arabic numerals; the rule of three; how to deal with the complicated metrological and monetary systems; alloying; partnership; simple and composite discount; the use of "single false position"; and area computation. Smaller towns might employ a master; in towns like Florence and Venice private *abbacus* schools could flourish. In both situations *abbacus* masters had to compete, either for communal positions or for the enrolment of students. (pp. 4–5)

Mathematical Activity and Mathematics Education

The term "mathematical activity" will be identified with making numerical calculations or measurements of quantity, or reasoning about relationships between physical or abstract quantities. Thus, for example, arithmetic, algebra, geometry, trigonometry, mensuration, calculus, and their direct applications (for example, in surveying or navigation), will be considered to be forms of pure or applied mathematics.

Mathematics education encompasses all matters concerned with the teaching or learning of mathematics, or with curricular issues arising from considerations of what mathematics might profitably be taught and learned by well-defined groups of people.

Nomenclature

In this book we shall use the term "cyphering book" rather than any of "ciphering book," "cypher book," "cipher book," "copybook," "copy book," "notebook," or "exercise book." In the 18th and early 19th centuries, in Great Britain and in North America, the three terms "cyphering book," "ciphering book," and "copybook" were often used to describe the same thing—sometimes the same person used the different terms interchangeably. We shall use "cyphering" in preference to "ciphering" because an "advanced Google search" revealed that during the period 1607–1820 "cyphering" (with a "y") was more commonly used in North America and in England than was "ciphering" (with an "i"). However, in 1788, Isaiah Thomas published an American version of William Perry's (1788) *Royal Standard English Dictionary*, and in this "cipher" was spelt with an "i." This continued in subsequent American editions of the *Royal Standard Dictionary*. In 1828, Noah Webster released his influential *American Dictionary of the English Language*, and in that book Webster gave "ciphering" priority over "cyphering." An "advanced Google search" revealed that after 1820 "ciphering" began to be more commonly used in published works than "cyphering." After about 1850, "cyphering" was rarely used.

The term "copybook" (or "copy book") suggests that everything was *copied* from another text. That was not always the case for manuscripts in our principal data set (PDS) described later in this book. Furthermore, the term copybook was also commonly used to describe handwritten manuscripts that dealt solely, or mainly, with non-mathematical content. The word "cyphering" implies the *writing of numerals*, which was the case in every manuscript in our PDS—including those mainly concerned with algebra, geometry, trigonometry, navigation or surveying.

Some writers (e.g., Williamson, 1928; Yeldham, 1936) have preferred the term "exercise book" to "cyphering book" or "ciphering book"—indeed, Yeldham, in her history of the teaching of arithmetic in Great Britain between 1535 and 1935, devoted a whole chapter to "the exercise book." However, we distinguish between cyphering books and exercise books for four reasons. First, in the 18th and 19th centuries "exercise book" was less used in North America than in Great Britain in relation to mathematical writing of students. Second, the term has subsequently been applied to manuscripts for any subject (and not just to mathematics manuscripts). Third, North American "cyphering books" were often larger in format than "exercise books"; also, their pages were usually unlined, whereas those in exercise books were usually lined. Fourth, exercise books used for mathematics were often manuscripts in which students wrote little more than solutions to problems set by the teacher. Since it was intended that cyphering books would become personal reference manuals, they contained definitions and statements of rules and cases, as well as problem statements and solutions.

We also distinguish between *abacus* boards and *abbacus* books. The former, which were physical devices on which reckoning was carried out (Howson, 1982), will not be the subject of much attention in this book. Although reckoning masters and students often used *abbacus* books and *abacus* boards in the same transactions, there was no connection, other than etymological, between the two (Franci, 2009; Van Egmond, 1980).

Chapter 2
Development of a Tradition

Abstract The cyphering tradition which controlled North American school mathematics between 1607 and 1840 was imported from Western European nations. The curriculum within the tradition had a strong mercantile emphasis, and was most suited to boys who wished to gain employment as clerks or reckoners, or as navigators or surveyors. This chapter traces how the cyphering tradition, which was originally developed in India and in Arab nations, was introduced into Italian city states around 1200. Gradually, it was adopted in Western Europe, including France, the Germanic states, Spain, Portugal, the Netherlands, and Great Britain.

Origins of the Cyphering Approach to Mathematics Education in Western Europe

In order to be in a position to understand the origins of the cyphering approach in mathematics education, one needs to appreciate how commercial, political, military and socio-cultural factors affected the development of numeracy in Western European nations and city-states or republics between 1200 and 1600. During that 400-year period, merchants increasingly exported and imported goods to and from distant, and hitherto unreachable, markets (Burke & Burke, 1758; Kretschmer, 1909; Long, McGee, & Stahl, 2009). Despite the ever-present danger of piracy and the difficulty of ensuring accurate navigation (Falola & Warnock, 2007), the increasing availability of faster and safer ships encouraged entrepreneurs to risk overseas venture capital (Howson, Keitel, & Kilpatrick, 1981; Swetz, 2005). The need for improved navigation practices spawned a group of "practical mathematicians" (Taylor, 1954), who invented magnetic compasses and other aids that helped speed up international commercial transactions (Collinder, 1954; Mahan, 1898).

The 17th century was the "golden age" in Dutch history, a time when the Netherlands was the world's greatest sea power. The Dutch East India Company (*Vereenigde Oostindishche Compagnie*, or VOC, in Dutch) was the first company in the world to issue stock, and by 1669 it had become the richest private company the world had ever seen. Not only did it have 150 merchant ships, 40 warships, 50,000 employees, and a private army of 10,000 soldiers, but, with respect to commercial activities it increasingly assumed many of the functions of the national government (Ames, 2008). Within that context, mathematics education became a vitally important matter, not only to the Netherlands, but also to competing commercial powers such as England.

Because the Dutch were instrumental in the early development of New York, Dutch coinage joined the Spanish, Portuguese, and British coins used for daily transactions in the American colonies. The arithmetic associated with currency exchange, and commercial investment, and the geometry associated with fast and accurate navigation, were matters of great importance for both Europe and America, especially for government authorities and merchants who had their eyes on local and national economic advantage (Kretschmer, 1909).

Merchants, keen to monitor and maximize profits, introduced new accounting and business procedures involving double entry bookkeeping, letters of credit, bills of exchange, and marine insurance (Long et al., 2009; Swetz, 1987, 2005). They increasingly looked for partnerships that would last for years rather than for a single transaction. Merchants began to specialize in what they did and in the materials they handled. Furthermore, the exploitation of cheap labor in Africa, the Caribbean Islands, and the Americas, and the use of indentured servants (often brought from England), promised to usher in an era of economic prosperity for the colonizers (Archer, 1988; Burke & Burke, 1758; Franci, 1992, 2009; Franci & Rigatelli, 1982, 1989; Long et al., 2009; Radford, 2003; Taylor, 1956; Van Egmond, 1980).

Eventually, prosperous merchant traders could be found in major West European cities—such as Florence, Genoa, Milan, Pisa, Venice, Geneva, Marseilles, Lisbon, London, Nuremberg, Leiden, and Amsterdam. Companies which needed to ship their wares to other cities and other nations developed central headquarters with networks of representatives along the principal trade routes. In the 13th century the principal trading city-states linked with Germany went so far as to create a semi-formal alliance of trading cities and guilds known as the Hanseatic League. This was established for the specific purpose of maintaining and extending a trade monopoly, and it remained in operation for several hundred years. Its primary purpose was to improve profits of member organizations by protecting trade routes, and coordinating commerce. The complex web of captains of business, accountants, shipping agents, other shipping personnel, distributors, labor-controllers, and manufacturers rapidly changed the face of international commerce (Kretschmer, 1909; Long, 2009).

Business leaders increasingly recognized that being able to calculate, predict, and control profit was important, and hence they sought out skilled "reckoners" who understood the new sophisticated arithmetic that had developed in relation to Hindu-Arabic numeration (Franci, 2009). These reckoners were expected to take advantage of place-value notation within the Hindu-Arabic numeration system and, in particular, to apply the four operations on whole numbers and fractions with currency exchange, weights, and other measures, discount, simple and compound interest, ratio and proportion tasks, and partnership calculations. The aim was to maximize the profits of those merchants who employed them.

As mathematics developed, the reckoners were expected to be responsible for harnessing its powers within areas of commerce involving *abbacus* concepts such as alligation, false position, fellowship, gauging, the rules of three, and tare and tret

(Arrighi, 1964, 1970; Long et al., 2009; Swetz, 1987, 2005). The Hanseatic League established a *Rechen Schulen* (a guild of reckoning masters) which, according to Brown and Coffman (1914), tried to keep instruction in business arithmetic in the hands of reckoning masters and out of any system of public schools that might be established.

The *Abbacus* Manuscript Tradition in Western Europe 1200–1600

The Emergence of the Italian *Abbaco* Tradition

The so-called *trattati* or *libri d'abbaco* were vernacular Italian pedagogic manuals of commercial mathematics, accounting, and geometry widely used in Italian reckoning schools (often called *scuole* or *bottegha d'abbaco*) during the period 1200–1600 (Bellhouse, 2005; Cowley, 1923; Karpinski, 1911, 1925, 1929; Long et al., 2009; Radford, 2003; Smith, 1908; Swetz, 1987, 1992). From the 13th century, sharp rises in international trade and banking prompted Italian city republics to form vernacular schools in which commercial mathematics, accounting and writing were taught to sons of merchants or to apprentices with important responsibilities.

In the city republics of Renaissance Italy it was a common practice for merchant-class parents to send their sons for two-year courses at "*abbacus* schools," where they learned practical, mostly commercial, mathematics known as *abbaco* (see, for example, Swetz, 1987, for details of such a text—the *Treviso Arithmetic*, an Italian arithmetic first printed in 1478 and translated into English by the American mathematics educator, David Eugene Smith). Several hundred handwritten Italian-language manuscripts associated with these schools, some prepared by masters and some by students, survive. Most of these deal only with arithmetic, but some also contain notes on algebra or advanced mathematical material. At least two of these *abbacus* manuscripts have been translated into English (Høyrup, 2007; Long et al., 2009). According to Swetz (1987), by the end of the 15th century, at least 30 *abbacus* texts had been printed in Europe (more than one-half of them in Latin, seven in Italian, four in German, and one in French). However, as Swetz (1987) also pointed out, at least one writer has claimed that before 1500 over 200 printed arithmetics had appeared in Italy alone.

Most problems in *abbacus* manuscripts related to commercial arrangements, payment of merchandise, commercial partnerships, weights and measurements, money exchange, etc. There were numerous tables in the manuscripts, some showing how distances could be calculated, and others illustrating different methods of accounting or of calculating interest (Long et al., 2009).

The *abbacus* approach to mathematics represented a radical departure from the humanist, classical educational curriculum in which mathematics was based on Roman numerals and was often limited to medieval Latin approaches. Critics maintained that this went little beyond the statement and illustration of rules for determining moveable feast days in the church calendar (Jackson, 1906), but others link that classical tradition to subsequent pure mathematics research.

Leonardo Pisano's *Liber Abbaci*. Although it is traditional to attribute the coming to Europe of the Hindu-Arabic numeration system and the *abbaci* to Leonardo Pisano, better known among mathematicians as "Fibonnaci," there is evidence that the system, and its notation, were imported to Europe from Arabic nations during the 10th century (Ifrah, 2000). Around 1200, Leonardo's father represented Pisa in a customs house in the Arab port of Bugia in Northern Africa. Leonardo grew up with his father in Bugia, and while there he studied under Arab reckoning masters. He traveled sufficiently widely within the African-Arab world to learn many of the Arabs' commercial and mathematical secrets. His best-known book, *Liber Abbaci,* was a massive compendium, written in Latin, which set out most of the known mathematical practices of the day. Its contents were drawn both from Leonardo's personal experiences and from a variety of Arabic texts on commercial arithmetic, algebra, and geometry. According to tradition, *Liber Abbaci* became a primary source of information from which the European world learned how the Arabic numeral system could be applied in day-by-day affairs (Gies & Gies, 1969).

The first section of Pisano's *Liber Abbaci* described the Hindu-Arabic numeral system, based on the numerals 0, 1, 2, 3, 4, 5, 6, 7, 8, and 9, and on a base 10 place-value numeration system that permitted large numbers to be represented compactly, and algorithms to be developed for addition, subtraction, multiplication and division. As would befit Pisano's links with the world of commerce, the second section of *Liber Abbaci* discussed examples from commerce through topics such as currency conversion, measurement of weights, distances and time, profit and loss, and interest calculations. In the third section, Pisano turned to more theoretical problems such as the Chinese remainder theorem, a classic rabbit-breeding problem that generated the famous Fibonnaci sequence, perfect numbers, Mersenne primes, arithmetic series, and square pyramidal numbers. The fourth section offered methods for obtaining rational approximations to irrational numbers. *Liber Abbaci* also included comments on Euclidean geometric proofs, and on methods of solving simultaneous linear equations (Gies & Gies, 1969).

Sasan Fayazmanesh (2006) has claimed that manuscripts by Fibonacci and a host of academic mathematicians who followed him were different from those written by merchants because "the merchants were interested in providing practical business training, particularly for their children" (p. 31). Although it is certainly true that Fibonacci did include recreational problems in his *Liber Abbaci,* it is also true that his main emphasis was on elaborating practical business arithmetic methods (e.g., in barter) as this was practiced by Arab and Indian merchants.

The structure of Fibonacci's *Liber Abbaci* provided a model for the numerous authors of printed "commercial arithmetics" that increasingly commanded attention in 16th-century Europe (see, for example, comments by William Davis (1960) and Humphrey Baker (1568) for the influence in France and in England, respectively, in the 16th century). Davis (1960), after pointing out that Baker (1568) dedicated his book to the "Company of Merchant Adventurers" in London, drew attention to the large influence of commercial arithmetics in Holland around 1600.

Van Egmond on the *Abbaci* and the History of Mathematics Education

Warren Van Egmond (1976, 1980) claimed that the *abbaci* represented a continuous and uniform tradition that stretched from the work of Pisano to the end of the 16th century, shifting from handwritten to printed books as the primary means of publication changed from manuscript to printing. Van Egmond (1980) argued that the tradition did not end there:

> The long list compiled by Augustus De Morgan in his *Arithmetical Books from the Invention of Printing to the Present Time* (London, 1847) shows that the *abbaci* are only the beginning of a tradition of arithmetical writing that extends to the present day. The elementary books that we all used to learn basic arithmetic are the direct descendants of the *abbacus* books of the fourteenth and fifteenth centuries. Many of the problems and methods that are first found here are repeated in our "modern" books. The *abbaci* are more than just a source for the history of mathematics and business in Renaissance Italy, they are themselves a distinct chapter in the history of mathematics. (p. 31)

Our analysis of various collections of handwritten European and North American cyphering manuscripts suggests that the handwritten and the commercially-printed *abbaci* provided the model for cyphering books in Western Europe and in other nations, including North America. However, once printed arithmetics became more available and affordable, the model evolved so that commercially-printed texts were often used as a complement to the traditional cyphering book. The order implied in the last sentence—that textbooks developed as a complement to the cyphering books—has usually been reversed by historians of education and commerce.

During the 14th and 15th centuries, Italian merchants were the business leaders of the Western world—and the *abbaco* tradition held sway in the counting houses of Continental Europe (Bellhouse, 2005; Long et al., 2009; Spiesser, 2004; Unger, 1888). Indeed, early in the 17th century, in the German city of Nuremberg alone, there were almost 50 reckoning schools in which students created *abbaci* (Swetz, 1987). Each nation had its own names for masters who either did the calculations needed for business transactions themselves or taught others how to do them and to record how exemplar problems were best solved. In Italy, the teachers were called *maestri d'abbaco,* in France, *maistre d'algorisme,* in Germanic regions, *rechenmeister,* and in England, "scrivener" (Howson, 1982; Ifrah, 2000; Swetz, 1987; Unger, 1888).

Many early *abbaci* featured a rhetorical genre, with problems written out in words. Thus, for example, a writer might state "divide 35 by 7 giving 5, and add 3 giving 8." Computations were *not* displayed, but were included within a single block of text so that they appeared in sentences. According to Van Egmond (1980), it was not until the end of the 15th century that *abbaci* writers began to display calculations. The period between 1200 and 1600 witnessed noticeable and practically significant changes in the arithmetic taught to students being trained to become reckoners. There was a definite move to make the content more directly relevant to commercial practices and to navigation (Long et al., 2009; Spiesser, 2004).

The major change, of course, was the introduction and standardization of the Hindu-Arabic place-value numeral system, but other common symbols such as the equal sign were introduced, and these changes revolutionized discourse related to business transactions. Developments in trigonometry came in the mid-1500s, decimal fractions were introduced around 1600, and logarithms soon after that. Not only would these developments profoundly influence the future of mathematics, but they also would affect mercantile practices, surveying, and navigation, in important ways (Brown & Coffman, 1914; Ifrah, 2000; Sarton, 1957; Smith, 1900).

Figure 2.1 shows a page from a 1599 German cyphering book. The top half of the page shows an example of a multiplication task. Then follows the heading "Division," and a "rule" which describes an algorithm for that operation.

Another feature of early *abbaci* was that detailed solutions to type, or model, problems were given immediately after the statements of problems. In almost all cases, a problem and its solution were written together in the same handwriting, a fact which contradicts speculation that teachers always wrote the problems and students solved them.

The *abbaci* were intended to serve as reference manuals. Some of them were specifically prepared to assist *abbaco* masters: these offered solutions to "type" problems to which masters could refer their students. A student would copy solutions into his own book, which might subsequently become an *abbacus* manual in its own right. It became a personalized, handwritten book to which the owner could refer later in life if and when he became a full-time practitioner. Although early *abbaci* were fundamentally books of problems and solutions, in time they also became statements of rules and cases. According to Bellhouse (2005), the writers of some *abbaci* attempted to solve well-known problems which had not been solved by anyone else.

Vera Sanford (1927) and Warren Van Egmond (1976, 1980) have provided comprehensive analyses of about 300 Italian *abbaci*. According to Raffaella Franci (2009), Maryvonne Spiesser (2004), and Van Egmond (1980), the most distinctive feature of *abbaci* was how Arabic numeration and algorithms were applied to the solution of word problems concerned with business contexts and practices. Thus, for example, Michael of Rhodes, a 15th-century Genevan mariner, typically solved a problem using several methods (Franci, 2009), and the various ways he went about obtaining his solutions would still be found in cyphering books 300 or 400 years later. Although solutions to problems other than business problems were found in some *abbaci*, without business problems a manuscript was not considered to be an *abbacus* manuscript (Franci, 2009; Van Egmond, 1980). A typical manuscript had between 100 and 250 pages of text, and each page presented and discussed problems that had numerical, and especially place-value, connotations.

The above summarizes the traditional thesis on the origins of the *abbaci*. An alternative thesis, one which has gained currency among some recent historians, is that the traditions derived from a more widespread culture of commercial mathematics, known to Fibonacci but first developed in Provençe or in Catalonia before reaching Italy (Heeffer, 2008; Høyrup, 2005a, 2005b, 2007; Van Egmond, 1976, 1988). Apparently, the first of these texts to introduce algebra was Jacopo da Firenze's

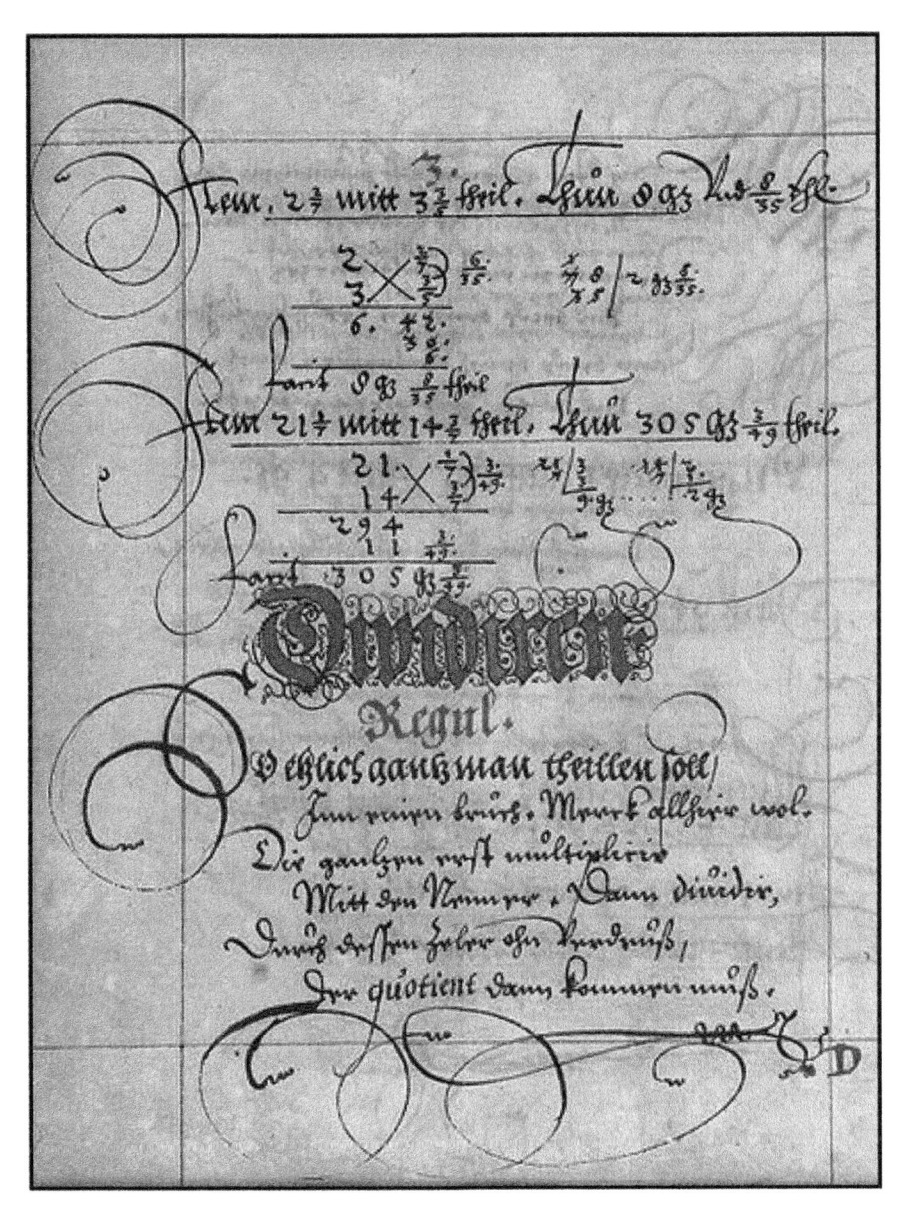

Figure 2.1. A page from the *abbacus* manuscript by Andréas Reinhard, 1599. (This manuscript is held at the Christianeum Gymnasium Historical Library, Hamburg.)

Tractatus algorismi, written in the early fourteenth century (Høyrup, 2005b, 2007; Smith, 1908; Swetz, 1987). That said, there are at least 15 extant 14th-century Italian manuscripts that include treatises on algebra (Franci & Rigatelli, 1988).

Whatever the origins of the *abbaci*, during the 14th, 15th and 16th centuries, *abbaco* schools flourished across northern Italy. Indeed, "approximately one-third to one-half of all Florentine boys during the late medieval and early Renaissance period may have attended school" (Emigh, 2002, p. 664), and many of the schools offered a specialist two-year course, for boys aged from 11 to 14, on mathematical techniques used in commercial enterprises (Benoit, 1988; Grendler, 1989). The tradition spread across Western Europe with, for example, Nicolas Chuquet, the early French mathematician, preparing a handwritten manuscript on commercial arithmetic (Fink, 1900). As an *escrivain*, Chuquet taught mercantile arithmetic on a daily basis to middle-class children in Lyon (Benoit, 1988; L'Huillier, 1976).

It would be wrong, however, to give the impression that the same content was to be found in cyphering books created in different parts of Western Europe. Paul Benoit (1988) has shown, for example, that the commercial techniques taught by Chuquet, on calculations of price, partnership contracts, divisions of profits and losses, and loans "were backward by comparison with those of the Florentines and the Venetians," but were "indicative of French commerce at the end of the Middle Ages" (p. 107).

An extant manuscript by Michael of Rhodes, a Genevan mariner of lowly social origins, can be definitely dated to the 1430s. It contains more than 180 pages of mathematics, all clearly in the *abbaco* tradition, with Arabic numeral calculations being performed in ink, on paper. Franci (2009) analyzed the methods used by Michael in the mathematics part of the manuscript, and concluded that Michael made full use of *abbacus* techniques associated with the four operations on numbers, the rule of three, alligation, fellowship, false position, barter, loss and gain, etc., to solve problems concerning the purchase and sale of goods (e.g., currency exchange, calculation and conversion of weights and measures). Michael also solved mathematics problems that were less obviously related to his daily life as a mariner, and often contrasted algebraic solutions with more numerical solutions (Franci, 2009; Long et al., 2009). He did this with obvious zest, and both Franci and Long were convinced that Michael loved mathematics.

The Content of the *Abbaci*

Typically, early *abbacus* problems had two components: first, there was a statement of the problem scenario, and second, a presentation and discussion, in the form of an "imaginary dialogue" of a recommended solution. Van Egmond (1980) regarded the following transcript which he (translated into English) as illustrative of this genre:

> A *soldus* of Provins is worth 40 *denari* of Pisa and a *soldus imperiali* is worth 32 *denari* of Pisa. Tell me how much will I have of these two monies mixed together for 200 *lire* of Pisa?
>
> Do it thus:
> Add together 40 and 32 making 72 [*denari*] which are 6 *soldi* and divide 200 *lire* by 6 which gives 33 *lire* and 6 *soldi* and 8 *denari,* and you will have this much of each of these two monies, that is 33 *lire* 6 *soldi* 8 *denari* for the said 200 *lire* of Pisa.
>
> And it has been done.
> Now, you are required to do all the following similar problems in the same way. (pp. 16–17)

Reckoning masters believed students learned best by copying type problems and their solutions, and by developing and writing solutions to associated exercises. Problems were carefully chosen so that if students themselves subsequently became reckoning masters they would have access to manuscripts that showed solutions to important classes of problems.

Van Egmond (1980) identified the following structure for a typical *abbacus* manuscript:

1. *Introduction.* After a discussion of the nature and value of numbers, the numeration system was described. Here the ten Hindu-Arabic numerals were shown, and the principle of place value explained, usually with an accompanying diagram. Then, the four main arithmetical operations on whole numbers were shown, and applied. Finally, fractions, and the key quantities—often involving conversions of monies, weights, and measures—were presented. Multiplication tables for numbers and money, tables of squares, and lists of parts of monetary units, would frequently be given.
2. *Business problems.* These ranged from simple calculations within one currency, to finding the price of a certain amount of a commodity by using a "rule of three," and to converting different currencies and weights and measures. Then came applied questions on loss and gain, discount, barter, fellowship (dividing profits between members of a partnership), simple and compound interest, equation of payments (when loans made over a period were combined for repayment on a single date), annuities, alligation (quantities of varying strengths were combined to make mixtures), and tare and tret.
3. *Recreational problems.* Van Egmond (1980, pp. 23–25) identified 12 types of problems in this category.
4. *Geometrical problems.* In this category, Van Egmond (1980) included three themes: the first was concerned with definitions; the second with abstract geometrical and measurement problems; and the third with applied measurement problems.
5. *Methodological problems.* Here, Van Egmond (1980) distinguished four subcategories: rule of three; rule of single false position; rule of double false position; and algebra.
6. *Miscellaneous material. Abbaci* often included lists of "tariffs" and other practical business information and variations in weights and measures found in the major trading city-states of western Europe and northern Africa. They quantified the gold and silver content of different coins in popular currencies, and stated sensible itineraries for those who regularly used major trading routes, as well as names and addresses of major banking institutions, and lists of major business events, etc. Astronomical and astrological information (including descriptions of the zodiac signs) were also likely to be included, as were calendars (including ecclesiastical holidays), medicines (for various types of illnesses), and literature (including poems, romances, chronologies).

According to Van Egmond (1980), the quality of handwritten *abbaci* varied considerably. The most elaborate of them featured exquisite penmanship, calligraphy, and colorful diagrams. These were almost always prepared by writing masters for immediate sale to wealthy merchants or to students from well-to-do families. They were usually well organized from a content point of view, but in some the calculations were scrappy, and even incorrect. Manuscripts prepared by apprentices tended to be less elaborate. Those who thought they might use the manuscripts they prepared later in their lives, tended to take more care, and put more effort into their books than students who, although they were receiving lessons from reckoning masters, never really intended to make large use of reckoning skills in the future.

Mathematics Beyond Reckoning

Reckoning was not the only part of mathematics that could assist enterprising business leaders to gain an edge. At the beginning of the 17th century, practitioners who could apply new navigational ideas involving the measurement of angles and distances on spherical surfaces were in great demand. Merchants knew that vessels plying their wares between Europe and the Americas, or the East Indies, needed mathematically-literate navigators who could guide them reliably towards distant specks on other continents, and could help them to arrive safely at required destinations in the least possible time (Mehl, 1968; Plumley, 1976; Schmidt, 1993).

The shortest path between two points on the earth's surface was known to be via a "great circle," but to be able to follow such a path required continual on-the-spot determinations of both latitude and longitude. Finding latitude was relatively easy, but the task of calculating longitude defied the world's top scientists for centuries. Governments of leading European sea-powers regarded the longitude problem as so important that they offered huge monetary rewards to anyone who could solve it—the Spanish government offered a prize in 1567, the Dutch in 1636, the British in 1714, and the French in 1715 (Dash, 2000; Taylor, 1956).

Meanwhile, it was recognized that each ship on the high seas should have at least one person, and preferably two or more persons, who could use mathematical instruments to locate the ship's approximate position, and could recommend the best way forward given the prevailing conditions (direction of the wind, etc.). Throughout the 17th and 18th centuries, ship-owners typically employed "naval schoolmasters" to instruct midshipmen, and others, during voyages, in reading, writing, arithmetic and navigation (Durkin, 1942; Taylor, 1966).

The Phillips Library in Salem, MA, holds a navigation cyphering manuscript completed in the 1770s by Henry Tiffin. From notes within the manuscript it is clear that Tiffin prepared this manuscript while serving as an apprentice midshipman aboard various sloops (especially the *Eagle* and the *Panther*). The mathematical content is concerned with straight-edge compass constructions, trigonometry, logarithms, profiles of coastlines, and various kinds of sailing (plain, oblique, Mercator, middle latitude). This manuscript includes numerous beautiful water-color works of art that Henry painted while on board the ship.

The Houghton Library at Harvard University holds two handwritten manuscripts (with 152 pages altogether) in which middle-level arithmetic was prepared in 1827 by a certain William F. Allen. According to a note on a front end paper in one of the books, the notebooks were sold by B. B. Macanulty, of Salem for use on the voyage "from Salem towards India on the *Barque Pompey*." The arithmetical content was standard (reduction, American currency, compound operations, decimal fractions, rules of three, inverse compound proportion, vulgar fractions, commission, brokerage, discount, insurance, barter, loss and gain, alligation, single position, American Customs duties), and probably represented the standard arithmetic curriculum for midshipmen going to India or the East Indies. After 1702, in England, those who wanted to be naval schoolmasters needed a "naval schoolmaster's certificate," and there is a record of a certain Nathan Prince, a Harvard College graduate and an experienced teacher of mathematics in Boston, going to London in the 1740s for the purpose of gaining that certificate (see, Taylor, 1966, pp. 139–140). A book, published in London in 1801, was specifically written for instructors of "sea youth" including "schoolmasters of the Royal Navy" (Morrice, 1801).

Heads of governments also began to recognize that mathematically competent navigators could help win wars in distant lands or on distant seas (Mahan, 1898; Waters, 1958). Newly discovered territories needed to be mapped, both internally and in relation to other territories and nations. Land needed to be systematically distributed and defended against hostile attacks. In short, the mathematical aspects of surveying, map-making, military strategy and fortification were important (Cohen, 1982). The emphasis on applied, as opposed to pure, mathematics would become particularly important in England, and would harden into a tradition that the British were more interested in developing instruments that would use the findings of mathematics than in developing the theoretical bases of the subject. In the 19th century, for example, the University of Cambridge would take great pride in the work of its applied mathematicians. By contrast, many Continental universities were more interested in developing mathematics departments which sought to prove theorems and develop pure mathematical theories (Curbera, 2009).

Changing Attitudes Toward the Value of Mathematical Studies

By 1670 the control of the American settlements on the east coast of North America had largely passed into the hands of England, and late-17th-century English politicians began to fear that their home education arrangements were not producing enough practical mathematicians to maintain their advantage (Howson, 1982). The relatively few English schools which provided an education that went beyond what was offered in the so-called "dame schools"—those small, private institutions that cared for children up to about the age of 10 (Cubberley, 1920; Monroe, 1917)—concentrated almost entirely on Latin grammar, with no branches of mathematics being taught. More seriously, there had developed an attitude among British genteel classes that the study of arithmetic should be confined to future clerks (Jones, 1954). Indeed, Eva Taylor (1954) claimed that around 1600 the upper classes in England regarded mathematics as worthless, something to be spoken of in the same breath as astrology and demonology.

The relative backwardness of England with respect to mathematics education had long been recognized (Howson, 1982). In 1557, for example, Robert Recorde (1557) lamented "the unfortunate condition of England, seeing so many great clerks to arise in sundry other parts of the world, and so few to appear in this our nation" (p. 212). Recorde went on to say that in England there was a "contempt," or disregard of learning, and that anyone who tried to teach mathematics to others was likely to be "derided and scorned, and so utterly discouraged to take in hand any like enterprise again" (p. 212). This attitude notwithstanding, by the end of the 16th century there had emerged a group of merchant adventurers in London who were vitally interested in arithmetic procedures of the *abbacus* variety (Baker, 1568, 1687; Davis, 1960).

Scholars like Lambert Jackson (1906) and Frank Swetz (1987) have claimed that during the 15th century there were four different types of mathematics vying for recognition within Western Europe: "theoretical tracts, *algorisms, abacus* arithmetic, and *computi*" (Swetz, 1987, p. 27). For Swetz, "theoretical tracts" were arithmetic derived from the writings of Anicius Manlius Severinus Boethius (c. 475–524), which focused on logical connections. That kind of mathematics held sway in the European universities as part of what was called the "quadrivium." In this arithmetic, numbers were often still represented by Roman numerals (Ifrah, 2000; Howson, 1982).

By contrast, the arithmetic based on *algorisms* (or algorithms) was concerned with applying the ten Hindu-Arabic numerals (0, 1, 2, ..., 9) with practical calculations—particularly in relation to business situations (Smith, 1908). *Abacus* arithmetic was the kind of arithmetic associated with the use of abacuses or other physical counting aids that were used at "counting tables," and *computi* were almanac-like calculations associated with calculations for determining dates of movable religious occasions (like Easter) (Reynolds, 1993; Smith, 1908; Swetz, 1987).

By the end of the 16th century the *algorism*-type of arithmetic, with its emphasis on Hindu-Arabic numeration and the use of standard procedures for addition, subtraction, multiplication and division for whole numbers, fractions, and compound numbers (i.e., the arithmetic associated with quantities such as time, weight, length, area, volume, etc.), had won the day among the business communities of Europe (Davis, 1960; Howson, 1982; Ifrah, 2000; Knox, 1788; Yeldham, 1936). Leading merchants recognized that it was profitable to employ competent arithmeticians of the *algorism* variety because they could use numbers to quantify what had happened, what was happening, and what might happen in virtually every facet of business.

The *algorism* practitioners separated all forms of business practice into discrete domains, and they claimed to have had a rule, and a case, for every situation. *Algorism* masters were expected to teach students important definitions, rules and cases and to show exemplary solutions to type problems by which the rules and cases could be arithmetically dealt with in applied contexts (Swetz, 1987). Employers sought out potential employees who had been trained by reputable reckoning masters in market-place arithmetic and who could represent them in their dealings with merchants and with reckoning masters. Therefore, at the beginning of the 17th century the demand for *algorism* practitioners increased rapidly right across Western Europe—even in nations as isolated as Iceland (Bjarnadóttir, 2009).

The "new rich" saw these practitioners as essential if they were to achieve steady improvement in the profitability of their businesses. Written arithmetics began to be printed, with authors obviously borrowing freely from previous authors, both in the forms of words used in explanatory text and in the enunciation of problems. Thus, for example, arithmetics by Sigismund Suevus, George Meichsner, and Leonhard Euler all provided data that enabled their readers to calculate the cost of building Solomon's temple, or the extent of the fortune of Sardanapaulus (the last king of Assyria) (Bjarnadóttir, 2009).

People with inherited property and influence in high society were not always convinced of the worth of this "new" mathematics (Davis, 1960; Ifrah, 2000; Swetz, 1987). For many of them, anyone who worked on or with mathematics was to be regarded as a peddler of the purely mechanical kind (Money, 1993; Taylor, 1954). Such attitudes were lurking within the world-views of religiously-oriented settlers, and even of some businessmen of English origins, who became part of the European communities in the far-flung North American settlements.

Linking Writing Education with Mathematics Education

The Cyphering Approach

In the 16th century, many children in Western European nations like Italy, Holland, France, and the Germanic regions, prepared cyphering books in community schools (Beaujouan, 1988; Benoit, 1988; Kool, 1988). Students in Calvinist or Lutheran communities were especially likely to be expected to "cypher" (Dewalt, 2006; Howson et al., 1981).

Klaas Van Berkel (1988) has shown that there were flourishing mathematical schools in the Dutch city of Leiden in the 16th and 17th centuries, and it is interesting that during the period 1580–1650 great mathematicians like Simon Stevin and René Descartes periodically lived in Leiden. Many of the New England "pilgrims" sojourned in Leiden before moving to Plymouth, Massachusetts, and although there is no known link between the pilgrims and mathematical scholarship in Leiden, arithmetic was valued throughout the city (Bangs, 2000; Campbell, 1892; Sarton, 1957). Apparently, in Leiden every child went to school, and almost every inhabitant could read, write and perform elementary calculations (Campbell, 1892). Such was the value that the Dutch placed on education that during the 17th century the Netherlands published more books than the rest of Europe combined (Campbell, 1892).

Marjolein Kool (1999) analyzed the content of thirty-six 15th- and 16th-century Dutch-language manuscripts dealing with arithmetic and concluded that students could use the Hindu-Arabic system of numeration not only to solve practical problems but also to explore mathematical concepts and ideas when tackling problems.

The importance that the Dutch and French Huguenot refugees who settled in New Netherland (now New York) would have attached to education is suggested by two woodcuts by Abraham Bosse (1602–1676), a Dutch master painter (McTighe, 1998). In his 1635 woodcut, *La Maitresse d'Ecole*, of a school in a Calvinist community in Holland, Bosse, who was from a Calvinist-Huguenot family, portrayed a spacious, simply-furnished room with orderly, well-dressed groups of young girls being attended to by an earnest female teacher seated at a table. Some of the girls had small books, and one was holding what appears to have been a hornbook. In a later (1638) woodcut, *Le Maitre d'Ecole*, Bosse showed a large room with 10 boys and 7 girls working studiously on various tasks. A seated male teacher, who was holding what appears to have been a cyphering book, was talking with a student. Two other boys, one holding a book that also looked like a cyphering book, were waiting their turn to speak with the teacher. Towards the back of the picture, seven boys were standing around a table probably preparing material to bring to the master for "recitation," or checking.

According to Louis Karpinski (1925), in Prussia and in Holland local *rechenmeisters* were "given a practical monopoly of the business of instruction in arithmetic" (p. 171). Dirk Struik (1936) maintained that in the 16th and 17th centuries the Dutch *rechenmeisters* supplied, in an increasing degree, trained personnel able to offer service in areas of commerce like accounting, computation, and navigation. Trades looked for practically-trained young men for careers in sea-faring, bookkeeping, cartography, astronomy, surveying, and the high demand forced Latin schools and Dutch universities to devote an increasing proportion of curriculum time to mathematics and physical and natural sciences.

An obvious gap existed between the Latinized university doctor and the "uncouth" teachers of arithmetic, the technicians and the book-keepers. Increasingly, classical institutions found themselves compromising their educational principles by offering mathematics and mathematics-related courses in which students prepared cyphering books in the local vernaculars (and not in Latin). Latin schools, in which the classics dominated the curriculum, would call in the *rechenmeister* to instruct students in basic cyphering (Littlefield, 1904). The *rechenmeister's* task was to pass on his professional knowledge to his students (Howson, 1982; Howson et al., 1981), and invariably he chose to do this by getting each student to construct a personal cyphering book.

In the 14th and 15th centuries, these cyphering books resembled the *abbaci* in content and genre and gradually the languages in which cyphering books were prepared changed from Latin to the local vernaculars. In time, the genre for cyphering books also changed (Struik, 1936), and the qualitative changes in genre that occurred will be considered later in this book.

The value that Dutch immigrants placed on arithmetic is evidenced by the fact that in New York in 1730 a Dutch-language textbook, *Arithmetica of Cyffer-Konst*, authored by a certain Pieter Venema, formerly a teacher in Holland, was published. A first edition of this text had been published in Holland in 1714, but for the New York edition, Venema undertook "to make a clear and succinct cyphering book upon that excellent science that flourishes in the city [i.e., New York] and country" (quoted in Carpenter, 1963, p. 126).

The content in Venema's text followed the *abbaco* tradition, dealing first with whole number operations and fractions, and then proceeding to "trade, partnership and the handling of merchandise" (Carpenter, 1963, p. 126). Translated from the Dutch, the title page of the book was "Arithmetic or the art of cyphering, according to the coins, measurements and weights of New York" (Simons, 1924, p. 59). But Venema also included a section on algebra because "that which is not understood in arithmetic can be demonstrated by the clear words of algebra" (quoted in Carpenter, 1963, p. 126). Simons (1924) stated that this was the first commercially-sold text including algebra to be written and printed in the American colonies.

The standard of handwriting and calligraphy in most cyphering books was high. That was partly due to the fact that many *rechenmeisters* doubled as writing masters. Venema (1730), for example, described himself as a writing master. In the 16th and 17th centuries in Europe, schools that instructed children in arithmetic were often known as writing schools, and so arithmetic instruction and writing instruction came to be intimately linked (Carlo, 2005). In England, only a small proportion of children studied arithmetic in schools, but those who did usually did so through a cyphering approach in which writing was emphasized (Heal, 1931; Monroe, 1917; Watson, 1902). Children were encouraged to practice their penmanship and calligraphy by making as elegant entries as possible into their cyphering books.

The link between arithmetic and writing continued throughout the 18th century and into the 19th century. In the German city of Nuremberg, for example, a guild united writing and arithmetic instructors, separating them from all other teachers (Cubberley, 1920). As Heal (1931), in his monumental study of writing masters and their copybooks, made clear, George Bickham, Edward Cocker, James Hodder, Humphry Johnson, and William Webster were among a large number of respected British writing masters who doubled as cyphering masters and authors of school arithmetics (Cajori, 1907; Carpenter, 1963; Freeman, 1913).

One consequence of this link between writing and arithmetic instruction was that when a school official stated that "writing" was part of the curriculum this often implied that students were expected to prepare cyphering books (Ayres, 1682; Bickham, 1740; Champion, 1747; Colson, 1736.; Jenkins, 1813; Johnson, 1719; Massey, 1763). The strength of this association has been lost in time, and this has resulted in the number of 17th- and 18th-century schools in which cyphering was an integral part of the curriculum being underestimated by historians (Small, 1914; Watson, 1902).

In England in the second half of the 17th century, in the 18th century, and for much of the 19th century, "writing as also arithmeticke" was recognized as important (Cremin, 1970; Heal, 1931; Monaghan, 2007; Reynolds, 1818; Smith, 2005). *Abbacus*-like texts by writing masters such as Humphry Johnson (1719), George Bickham (1740) and William Butler (1788) featured exquisite calligraphy (see for example, the copy of a title page shown in Figure 2.2).

In Butler's (1788) text, exercises were beautifully presented and spaces were left for students to write in answers. Despite its title ("An Introduction to Arithmetic Designed for the Use of Young Ladies"), Butler's (1788) text was written primarily to appeal to arithmetic instructors. On his title page, Butler described himself as

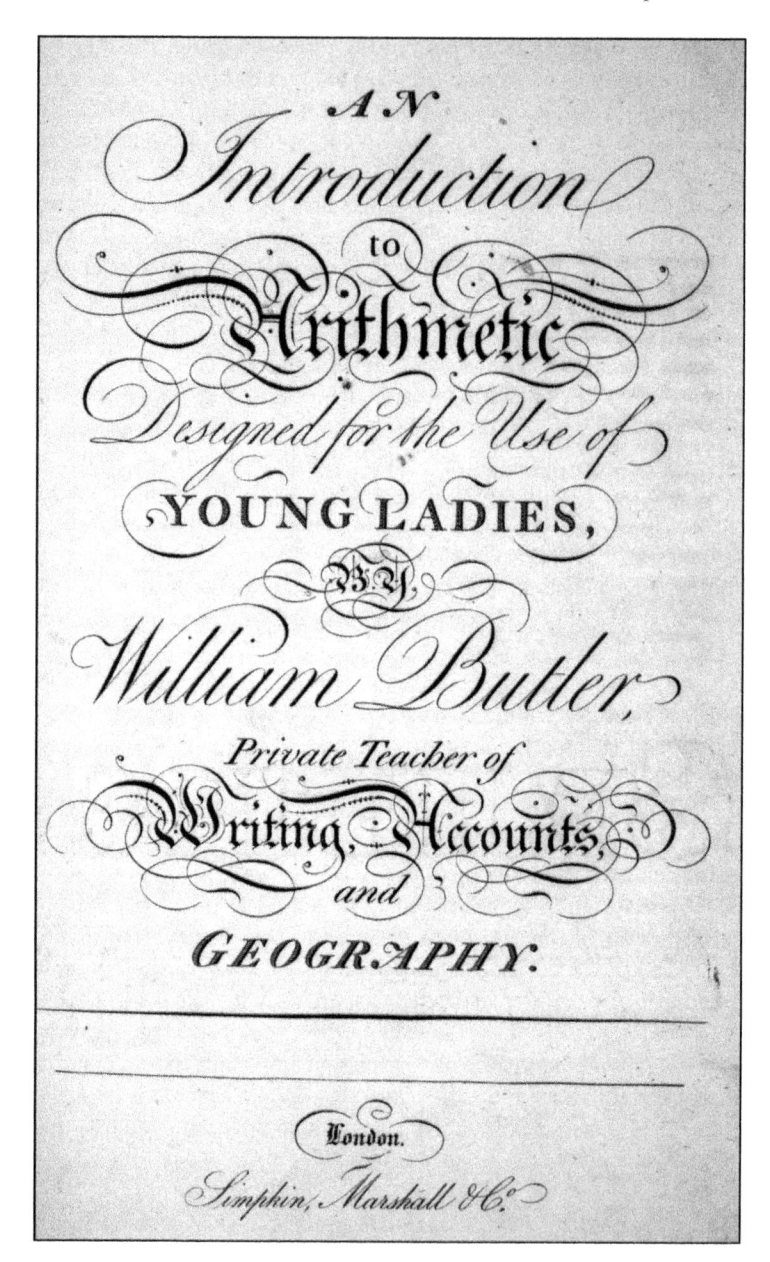

Figure 2.2. The title page of William Butler's (1819), *An Introduction to Arithmetic: Designed for the Use of Young Ladies* (London: Simpkin, Marshall & Co.).

a "private teacher of writing, accounts and geography." In the preface to the 1819 edition of his book, Butler stated that he had been flattered by the appearance of other publications which obviously mimicked his first edition. He claimed that "of these, seven or eight have appeared in London, and two in the country" (p. iii).

The Role of Mathematics in Education

Many community-based schools in Reformation and post-Reformation Europe had to meet standards set by outside authorities. In addition to learning to count, most school children were expected to perform written calculations and to solve problems, in writing, that required computations. In 1548, for example, the Bavarian *Schuelirdnungh* introduced arithmetic as a required study in village schools in Bavaria (Fink, 1900).

In 1642, Duke Ernst the Pious of Gotha, in Prussia, issued a regulation for schools in his duchy that not only made schooling compulsory for all children aged between 5 and 12, but also defined a curriculum which included arithmetic as a required subject. The school year would occupy 10 months, and children would be compelled to attend every day, from 9 am to 12 noon, and from 1 pm to 4 pm. According to Monroe (1922), this became the standard for almost all parts of Prussia.

Its practicality notwithstanding, in many parts of Europe, arithmetic was not seen as the kind of study in which children from well-connected families should engage. Mathematics of a purist, fundamental kind might be deemed a socially acceptable area of study—although that was not to be taken for granted—but arithmetic, navigation and surveying, which were aimed at achieving practical ends, were deemed by some to be of questionable worth for the children of those in high society (Fink, 1900; Recorde, 1549; Thut, 1957).

The period 1300–1700 was one in which European scholars like Cardano, Chuquet, Desargues, Descartes, Fermat, Galileo, Harriot, Huygens, Kepler, Leibniz, Napier, Newton, Oughtred, Pacioli, Pascal, Ptolemy, Regiomontanus and Stevin made lasting contributions to the development of modern mathematics (Fink, 1900; Hay, 1988; Smith, 1908; Van Egmond, 1988; Whitrow, 1988). Given the outstanding individual mathematical achievements across Western Europe, one might imagine that the systems of mathematics education which generated these scholars would be especially worthy of study in the New World. Furthermore, since by the middle of the 17th century most of the eastern North American settlements had come to be under the control of England, the situation in England might be expected to be of particular interest (Howson, 1982).

It was matter of dispute, however, whether the kinds of algorithmic procedures, and the content, associated with the cyphering tradition, were genuinely educational. For, had not Plato declared that arithmetic "had value as a study only if pursued in the spirit of a philosopher and not of a shopkeeper" (Thut, 1957, p. 63)? What constituted mathematics, and what kind of mathematics might be worthy of study in schools and colleges, were issues that were never far below the surface of higher education. But such matters were of little, if any, interest to the merchants who wanted to maximize profits (Ifrah, 2000).

In England in 1677, John Newton, a strong advocate for the introduction of more mathematics into the curricula of grammar schools (Taylor, 1954), asserted that he did not know of any grammar school in England in which mathematics was taught (Jones, 1954). That seemed to contradict a 1612 Charterhouse School statute which had decreed that the school-master and the usher should "teach the scholars to cipher and cast an accompt, especially those that are less capable of learning, and fittest to put to trades" (A Carthusian, 1847, p. 127). The grammar schools concentrated on classics, in order to meet the needs of prospective clergymen, lawyers, and physicians (Cremin, 1970, 1977; Howson, 1982; Howson et al., 1981; Littlefield, 1904).

Although the élite English schools may not have always included arithmetic in their curricula during the first half of the 17th century, there can be no doubt that cyphering books were often prepared in England at that time. For example, Thomas Dixon prepared a 300-page cyphering book around 1630 which featured exquisite calligraphy and penmanship. The language used was English and not Latin, and the practical topics recorded, and the genre in which they were set out, were unquestionably consistent with the *abbaco* tradition. We inspected this cyphering book at the Butler Library, Columbia University, where it is part of the Plimpton Collection Post 1600 (MS 510, 1630).

In fact, during the 17th and 18th centuries the cyphering-book tradition gradually took hold in England, especially in the dissenting academies (Hans, 1951a; Parker, 1914) and at privately-operated writing and reckoning schools (Parker, 1912; Watson, 1902). Sometimes, students copied entries into these cyphering books from textbooks (Cockburn, King, & McDonnell, 1969). Thus, for example, during much of the first half of the 18th century Robert Hartwell, of London, offered instruction in "arithmetic, geometry, astronomy, cosmography, geography, navigation, architecture, fortification, horologiography, &c," while attending to the doctrine of triangles, "plaine and sphericall," and to "the use of the tables of sines, tangents, secants and logarithms" as well as to "accompts for merchants by order of debtor and creditor" (quoted in Karpinski, 1925, p. 172). Hartwell became well known as a latter-day editor of Robert Recorde's *Ground of Artes* (Howson, 1982), and almost certainly his students would have used their master's edition of Recorde's book as the basis for entries in their cyphering books. However, although not all of his students would have been able to afford to purchase the textbook itself, they would have been able to take their cyphering books with them and use them as personal reference books for the rest of their lives.

John Newton's 1677 statement notwithstanding, then, there is evidence that there were significant mathematics education developments in England in the 17th century. We shall argue, however, that the special role of one particular school, the Royal Mathematical School at Christ's Hospital in London, came to be of central importance.

The Royal Mathematical School at Christ's Hospital, London

Perhaps the finest cyphering book in our personal collection of over 260 cyphering books was produced by a certain Charles Page, who attended the Royal Mathematical School at Christ's Hospital, in London, between 1818 and 1825 (Christ's Hospital admission register). According to Jones (1954), Christ's Hospital was the first English school in which mathematics became an important part of the curriculum. When Alexander Dallas Bache (1839), Benjamin Franklin's grand-son, visited Christ's Hospital in the 1830s, he declared, that its grand scale was "so much beyond any other in Great Britain, and, indeed, in the world" (pp. 65–66). Hans (1951b) described the Royal Mathematical School at Christ's Hospital as the first "modern school" established anywhere in the world (p. 532).

In 1673, King Charles II and his advisers, recognizing that sea-power was becoming crucially important—not only for national defense and status, but also for colonial expansion and for mercantile and industrial development (Howson, 1982; Howson et al., 1981)—started a mathematical section at Christ's Hospital for 40 scholarship "orphan" boys aged from about 10 years. These students were supported out of royal finances, and the implementation of the concept was supervised by a board which included the nation's top scientists (Christ's Hospital, 1953; Howson, 1982; Pearce, 1901; Plumley, 1976).

The decision to establish the Royal Mathematical School was undoubtedly motivated by a desire to keep up with the Dutch for, in 1600, Maurits, Prince of Orange and Stadholder of Holland and Zeeland had founded a school for military engineers in Leiden (Waters, 1958). The statutes and the curriculum for this school had been drawn up by Simon Stevin, a famous mathematician, and lessons were given in the vernacular, rather than in Latin. The school was intended for "illiterate" masons, navigators, carpenters, surveyors and engineers, and after attending standard "theoretical" classes in the winter, selected students were conscripted to the Dutch army to "apply" what they had learned. This new school was not a university, but during the 17th century it profoundly influenced mathematics curricula in Dutch universities (Van Berkel, 1988).

In the 1670s it was planned that, from the outset, the Royal Mathematical School at Christ's Hospital, in London, would bolster the theoretical and practical backgrounds of recruits for the British merchants and for the British navy (Hans, 1951a, 1951b). The early mathematics masters struggled to cope with the demands of the curriculum and teaching, however, and it was 35 years before the Christ's Hospital board was able to find someone—James Hodgson—sufficiently competent mathematically and pedagogically to cope with the demands of office. Hodgson would be Master of the Royal Mathematical School for 46 years (Christ's Hospital, 1953; Plumley, 1976).

The Royal Mathematical School in Christ's Hospital was expected to provide the best possible training in mathematics, especially as that subject related to navigation and surveying (Howson, 1982). The school developed a curriculum and an approach to instruction that relied heavily on the complementary aspects of two main inputs: first, a comprehensive commercially-printed text prepared by a master of the school;

and second, a cyphering approach. The curriculum was carefully thought through, with Isaac Newton having considerable influence (Allan, 1984). Newton was determined that it go well beyond the level of mathematics normally expected "in the vulgar road of seamen," so that in time the school would "furnish the nation with a more skifull sort of sailors, builders of ships, architects, engineers and mathematicall artists of all sorts, both by sea and land, than France can at present boast of" (quoted in Turnbull, 2008, p. 358, Newton's original spelling retained).

Every student was expected to prepare a handwritten record of what he studied, and the cyphering book that resulted was to be used when the student was apprenticed as a navigator or reckoner for a private merchant (Allan, 1984; Hans, 1951b). Royal Mathematical School graduates were required to serve as apprentices for seven years. Although they did not usually proceed directly to a university (Bache, 1839), there were exceptions (Brooks, 2010).

The boys in the Royal Mathematical School were expected to maintain the highest possible standards of penmanship, calligraphy and illustrations in their cyphering books in which details of the arithmetic, geometry, trigonometry, and navigation components of the curriculum were meticulously recorded (Howson, 1982). The appointment of Master to the Royal Mathematical School was highly sought after, and even persons nominated by Isaac Newton were not necessarily appointed (Trollope, 1834). Christ's Hospital masters knew that they carried the burden of responsibility for preparing students who would be capable of guiding large merchant ships, or even warships, across the Atlantic Ocean to the Americas, or around the Cape of Good Hope to the East Indies.

Christ's Hospital would become one of the world's best training schools for navigators and by the beginning of the 19th century Great Britain was unquestionably the leading maritime power in the world (Mahan, 1898; Padfield, 2005; Waters, 1958). As early as 1697, Peter the Great, Czar of Russia showed interest in the school and two young Christ's Hospital graduates (one was 15 years old, the other 17) were appointed to lead a training school for navigators that opened in 1701 in Russia (Allan, 1984; Cross, 2007; Hans, 1951b; Howson, 1982). Graduates of the Moscow School of Mathematics and Navigation would subsequently become teachers in the famous cyphering schools that Peter the Great established in 1716 to teach arithmetic and geometry to children drawn from all levels of Russian society. The schools were located in different provinces across the nation (Cracraft, 1971; Okenfuss, 1973).

Some masters of Christ's Hospital's Royal Mathematical School were of the highest order, both as mathematical scholars and as mariners (Hans, 1951b). For example, William Wales (1734–1798), the highly respected mathematician who was senior master between 1775 and 1798, was a Fellow of the Royal Society. Between 1772 and 1775 Wales was co-navigator with Captain James Cook on Cook's second major journey (Christ's Hospital, 1953; Pearce, 1901; Wilson, 1842).

The influence of Christ's Hospital would subsequently be transplanted to many parts of the world because, after completing their apprenticeships, some graduates became schoolmasters, navigators and surveyors in the New World (Allan, 1984; Brooks, 2010; Coldham, 1990). On finding that four or five years were not enough to train competent navigators, Hodgson arranged for royal scholarships to be extended so that some students remained at school for up to 10 years (Christ's Hospital, 1953; Trollope, 1834; Wilson, 1842), but after Hodgson's departure, the original intention for boys to be aged between 10 and about 16 was re-established. The boys received training in writing, arithmetic, algebra, geometry, trigonometry, surveying, astronomy, navigation and seamanship, and the mathematics standards set at Christ's Hospital were high. Figure 2.3 shows a page from a cyphering book created

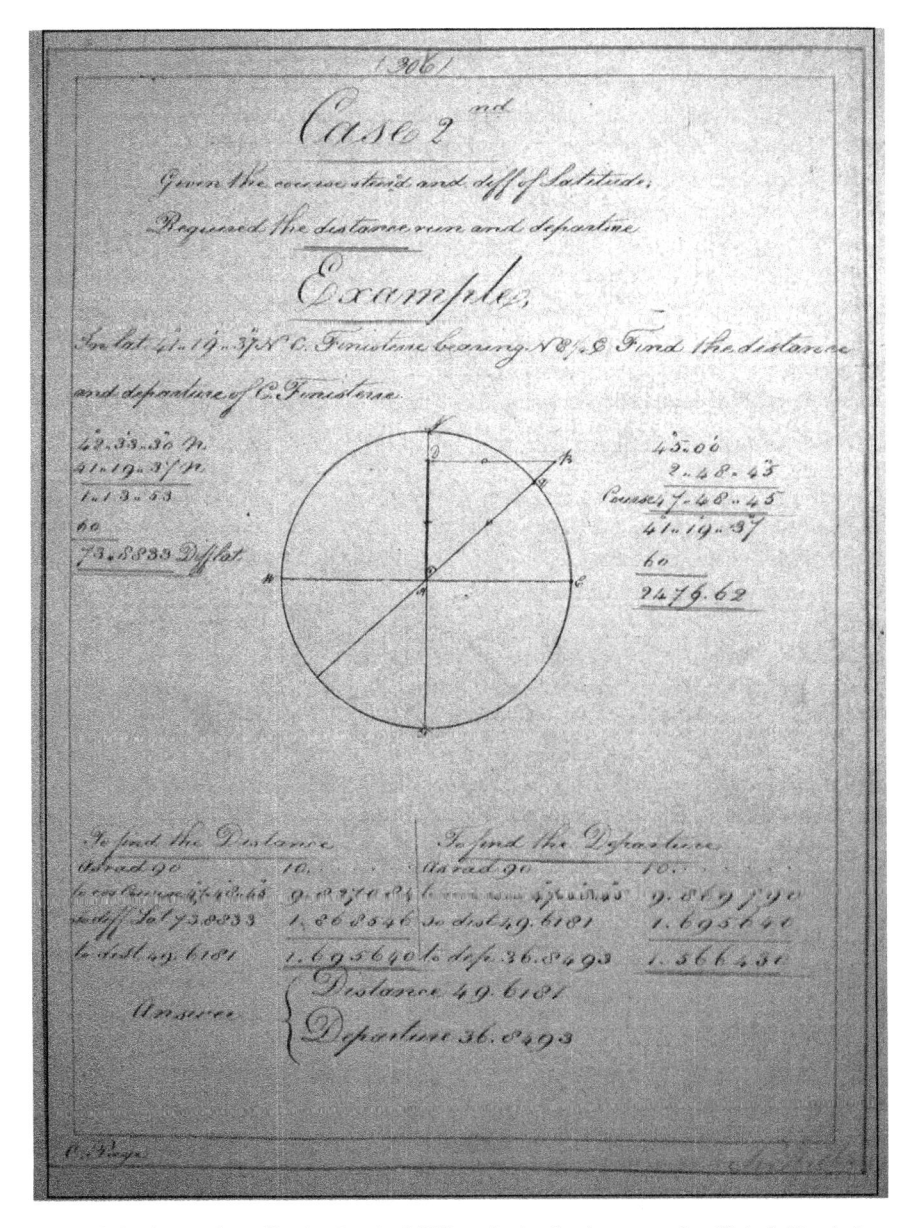

Figure 2.3. A page from Charles Page's (1825) cyphering book, prepared at Christ's Hospital.

by Charles Page, a student at Christ's Hospital between 1818 and 1825. Considering that Page was aged between 8 and 15 years when he prepared his cyphering book, the quality of his penmanship was remarkable.

Page's book, which has 671 handwritten pages, each with dimensions 11.5″ by 9.5″, is divided into 14 sections: Arithmetic, geography, the *Elements of Euclid* Books 1, 2, 3, 5 and 6, plane trigonometry, chronology, algebra, navigation, oblique sailing, globular sailing, spherics, astronomy, the use of globes, day's works, and "journal of a voyage from England towards Madeira." The last section was based on an imagined journey, for students were rarely sent on *actual* journeys. The standard of penmanship and calligraphy throughout Page's manuscript was high, and the manuscript was probably something that Page proudly owned, and used for the rest of his life.

We have examined the cyphering books of seven other Christ's Hospital students—prepared by James Poad Anderson in 1840, Henry Josiah Calkin Andrew in 1836, James Janeway in 1767, Andrew Cook Mott in 1837, Benjamin Raffles in 1755, Thomas Welton Stokoe in 1790, and James West in 1785. All of them featured extraordinary penmanship and calligraphy, and an impressively high level of mathematics. Details relating to these and other students can be found in handwritten school registers (e.g., Christ's Hospital, 1857). We own a 697-page published treatise on mathematics and navigation, *A System of the Mathematics, Containing the Euclidean Geometry, Plain and Spherical Geometry, the Projection of the Sphere, both Orthographic and Stereographic, Astronomy, the Use of Globes and Navigation,* which was the first of two volumes that James Hodgson (1723) specially prepared for students at Christ's Hospital.

The idea that cyphering at Christ's Hospital should be based on a curriculum derived from high-quality written mathematics textbooks had been implemented from the opening of the Royal Mathematical School in the 1670s, and a succession of textbooks written by Christ's Hospital mathematics masters were prepared. Thus, for example, in the 1750s, the boys in the Royal Mathematical School based their cyphering on *The Elements of Navigation Containing the Theory and Practice, with the Necessary Tables and Compendiums for Finding the Latitude and Longitude as Sea, to which is Added a Treatise on Marine Fortification,* the first edition of which had been prepared by John Robertson, Master at Christ's Hospital and later at the Royal Naval Academy in Portsmouth. Robertson's text had clearly strong similarities with the text, written by James Hodgson (1723). Later, Robertson's text was modified by the new senior Master of the Royal Mathematical School, William Wales (see Robertson & Wales, 1796).

Although all Royal Mathematical School students had access to detailed descriptions of the mathematics that they had to learn in the written texts provided by the School, each student was still expected to prepare a cyphering book that would be his book "for life." In fact, the mathematics curriculum at Christ's Hospital Royal Mathematical School did not vary much throughout the 18th century (Hans, 1951b). The standard of mathematics was high and the standard of penmanship in the boys' cyphering books equally high. After 1733, a similar curriculum was followed at the Royal Naval Academy at Portsmouth. John Robertson, mathematics master at Christ's Hospital, became mathematics master at Portsmouth in 1755, and in time the quality of cyphering books prepared at that institution rivalled the quality for Christ's Hospital Royal Mathematical School (Bache, 1839; Howson, 1982).

"Writing as also Arithmeticke" at Christ's Hospital

In 1577, Christ's Hospital established an endowed writing school (Freeman, 1913), and immediately that school included cyphering as a key component of its curriculum (Christ's Hospital, 1595). When, almost a century later, the Royal Mathematical School was established, school authorities immediately demanded high excellence in the scholars' penmanship (Howson, 1982; Trollope, 1834). Thus, "writing as also arithmeticke," already an important component of the European *abbaco* tradition, became an essential aspect of the Royal Mathematical School's program.

This emphasis was consistent with prevailing beliefs that not only was high-quality penmanship a sign of a well-educated person, but also that learning to write numerals and to set out calculations attractively were expected outcomes of writing education. The three most important English writing instructors of the 17th century, John Ayres, James Hodder and Edward Cocker, each taught "writing as also arithmeticke" (Karpinski, 1925, p. 173; Thornton, 1996). Samuel Pepys (1995), the famous English diarist of the late 17th century, confidant of British royalty and Fellow of the Royal Society (Howson, 1982), described Cocker as "the famous writing master" and after consulting Cocker in 1664, he recorded in his diary how impressed he was with Cocker's work. In that same year, Cocker (1664) had published a 56-page book aimed at those who would tutor in writing and arithmetic. That book dedicated just seven pages to writing, but 49 to arithmetic. In 1671, Edward Wingate, the author of another famous textbook on arithmetic, had published *The Clark's Tutor for Arithmetick and Writing, or a Plaine and Easie way of Arithmetick.*

The high quality of presentation in the Christ's Hospital cyphering books continued until well into the 19th century (Reynolds, 1818). In 1843, more than 150 years after Samuel Pepys had insisted on the highest standards of penmanship in cyphering books at Christ's Hospital, Horace Mann the well-known American educator, visited the school and reported that the students' writing there was "the most beautiful" he had ever seen (Mann, 1937, p. 185).

The "writing-as-also-arithmeticke" culture that permeated the Royal Mathematical School at Christ's Hospital was part of a carefully conceived mathematics education policy developed by an advisory body of scientists appointed by the British government (Howson, 1982; Trollope, 1834). Christ's Hospital students quickly learned that only their best penmanship and calligraphy would suffice when they were making entries into their cyphering books. They also knew that their cyphering books would become an instrument for evaluation, in the sense that they would have to present them to the School's annual examiners, and perhaps even to the British King. Stamped in gilt on the front and rear leather covers of the cyphering books by Charles Page and Henry Josiah Calkin Andrew—prepared in 1825 and 1836 respectively—is the royal crest. We own both of those manuscripts.

The Christ's Hospital Model for Mathematics Education

As we pointed out earlier, the Christ's Hospital mathematics program was devised by a committee that included people of the intellectual caliber of Jonas Moore, Isaac Newton, Samuel Pepys, Peter Perkins, John Wallis and Christopher Wren. Despite occasional modifications, it would remain in place for at least 150 years (Howson, 1982; Trollope, 1834). Curiously, Geoffrey Howson (1982), in his otherwise illuminating analysis of Samuel Pepys' work in supervising the development of the Christ's Hospital Royal Mathematical School's curriculum in its early years, did not acknowledge that from the outset the cyphering approach, complemented by a high-level textbook especially written for the Christ's Hospital program, defined the *modus operandum* for the teaching of mathematics.

The early teachers in the Royal Mathematical School at Christ's Hospital struggled to cope with the demands of their appointments (Allen, 1970; Howson, 1982; Pearce, 1901). However, the program survived as a result of expectations that each scholar would copy, in the best possible penmanship and calligraphy, chapters in a well-focused text (written by Jonas Moore and Peter Perkins (1681), both members of the Royal Society). In the Moore and Perkins (1681) textbook, Moore wrote chapters on arithmetic, geometry, trigonometry and cosmography, and Perkins, chapters on algebra Euclid, and navigation. This suggests that around 1700 mathematicians and authorities at the highest level of government in England were convinced that the cyphering approach, complemented by authoritative texts, provided the most effective form of mathematics education possible.

At a time when sea-power was important for determining international status, because of its relevance to trading and commerce, to war, and to the establishment and securing of far-flung colonies, the English government moved to support other specialist schools that would train personnel for the Royal Navy (Allen, 1970; Howson, 1982; Mahan, 1898; Taylor, 1956; Trollope, 1834). Navigators needed reliable, on-the-spot guidance for problem solving, and the handwritten journals (we shall call them "navigation cyphering books") prepared by students attending the Christ's Hospital Royal Mathematical School provided tangible models from which other institutions developed programs based on similar education philosophies.

As the 18th century progressed, numerous schools (e.g., Greenwich Hospital, the Grey Coat School at Westminster, the Plymouth Workhouse), and private tutors, in Great Britain, began to specialize in mathematics (Allen, 1970; Hans, 1951a; Howson, 1982). In 1722, Great Britain established its Royal Naval Academy, and in 1741 its Royal Military Academy (Howson, 1982; Howson et al., 1981), and these adopted the cyphering approach for training students in advanced arithmetic, algebra, geometry and trigonometry, surveying, and navigation (Howson, 1982; Karpinski, 1925; Watson, 1913).

It seems that this influence not only quickly spread across Great Britain (Bache, 1839), but it was also translated into the American colonies. We have examined cyphering books originally prepared at Christ's Hospital or at the Royal Naval Academy in the 18th century but now held in the Clements Library at the University of Michigan, in the Houghton Library at Harvard University, in the Phillips Library at Peabody Essex Museum, Salem (Massachusetts), in the American University, Washington, DC, in the New York Public Library, and in the Butler Library at Columbia University, New York. In 2011, we located and examined three original Christ's Hospital cyphering books held in the London Metropolitan Archives and in Guildhall, London—to date, the only cyphering books to be found in the school's records.

Great Britain was not the only nation to take steps to increase the number of persons with training in applied mathematics available for government service (Allen, 1970; Howson, 1982; Schubring, 2005; Taton, 1986). In 1671, for example, the French government began to train future French army officers in royal artillery schools. In 1720, the government founded *écoles régimentaires d'artillerie* for the special training of five garrisons of artillery regiments (Schubring, 2005), and by 1789 there were eight such schools, each using a cyphering approach complemented by specially prepared textbooks (Herttenstein, 1737; Taton, 1986). These schools taught applied arithmetic, geometry, trigonometry, mechanics, fortification, and civil architecture (Griffith, 1976). France's *Ecole Polytechnique*, established in 1794, represented a natural development. France also established impressive navigation schools (Allen, 1970). In the United States, West Point Military Academy was established in 1802 (Howson et al., 1981).

In 18th-century Great Britain, "writing as also arithmeticke" was not confined to schools preparing naval apprentices (Monaghan, 2007; Smith, 2005). Until well into the second half of the 19th century, an important aim of many private tutors and schools in Great Britain was to assist students to prepare *abbacus*-like cyphering books. That was made clear in *abbacus*-like books by well-known writing masters, Humphry Johnson (1719), George Bickham (1740) and William Butler (1819), whose own texts featured exquisite calligraphy. Exercises were beautifully presented, with spaces being left for students to write in answers. Charles Hutton, who would become a well-regarded mathematician (Howson, 1982), once described himself as a "writing master and mathematics teacher" (Hutton, 1766, p. i). Hutton argued that since "all mercantile affairs as well as all arts and sciences do absolutely depend on arithmetic" (p. ii), all scholars should prepare a printed book in which they recorded their work in arithmetic. This was reminiscent of Isaac Newton's view that elementary arithmetic provided the basis for all higher studies of mathematics (Pearce, 1901).

Toward a Theoretical Model for the Cyphering Approach

Bickham and Butler were not the first writers to prepare arithmetics in which students were invited to write answers. In North America, for example, Isaac Greenwood (1729), the first Hollis Professor of Mathematics at Harvard College, prepared such a book. Later, Daniel Adams (1801, 1822) prepared widely-used books which adopted the same approach.

The cyphering approach adopted in North American settlements in the 17th, 18th and 19th centuries derived from the *abbaco* tradition. Van Egmond's (1980) analyses have revealed that *abbaci* documents were of two kinds—specifically, handwritten books prepared by writing masters for the use of reckoning masters and the children of the well-to-do, and handwritten books prepared by students in reckoning schools (Unger, 1888).

The former were concerned with definitions and rules relating to carefully chosen type problems, and with careful statements of problems and their solutions. Occasionally, there were also exercises for students to solve, but if there were, solutions to these exercises were not provided. The handwritten books prepared by students included the type problems with their solutions *and* the exercises *with* their solutions. The aim was for students to learn to solve exercises for each type, so that they would be able to solve them in later life.

Abbaci books and manuscripts often included a set of "promiscuous problems" or "miscellaneous questions." Wingate (1630), writing in the context of vulgar fractions, referred to them as "a collection of subtil questions to exercise all the parts of vulgar arithmetic" (p. xii). The problems in these sets belonged to major classes of problems that had been dealt with earlier in the book. A student's task was to decide, for each problem, which major class, which sub-class, etc., was relevant, and then to solve the problem by the "best" method.

Sets of "promiscuous" problems (e.g., Bennett, 1815; Butler, 1819; Hutton, 1766; Pike, 1829), sometimes called "miscellaneous problems" (e.g., Mann & Chase, 1851, pp. 327–376; Sterry & Sterry, 1795, pp. 115–117), found in many arithmetics used in the 18th and 19th centuries, were intended to serve the same purpose. Almost one-half of the cyphering books in the PDS for this investigation included statements of, and solutions to, "promiscuous problems."

A Structure-Based, Problem-Solving Model Implicit in the Cyphering Approach

Reckoning masters following the *abbaco* tradition required students to follow a sequence that depended on model problems and their solutions being recorded accurately in the students' own handwriting. The masters believed it was their principal task to help their students recognize the general arithmetical structures to which particular problems belonged, and to assist the students to develop handwritten records of problem-solving methods that they could use for the rest of their lives (Ellerton & Clements, 2009b). Each major problem category had sub-categories ("cases"), and for each sub-category a type problem was chosen and solved. Students were then asked to solve problems that were structurally identical to the type problems.

It was important that everything written in a cyphering book be totally correct, for the book was to become the writer's personal guide for later life. It was assumed that the act of writing down correct solutions to problems compelled students to "reflect" on the structures of the problems, and made them more likely, in the future, to be able to identify problems embodying such structures (Heal, 1931; Monaghan, 2007; Van Egmond, 1976).

As those responsible for the Christ's Hospital mathematics program quickly learned (Allen, 1970; Trollope, 1834), for such an approach to be maximally efficient there were at least four important education requirements:

1. A teacher needed to know enough mathematics to be able to help students when they required it. This requirement related to the teacher's *mathematical* knowledge.
2. The teacher needed to be capable of communicating corrective feedback. This related to the teacher's *pedagogical* skills.
3. For their part, students needed to know enough mathematics to be able to cope with what they were being asked to do. This related to *student readiness*.
4. There needed to be sufficient time and facilities available in order that a student would be able to solve set problems. This was the *resource* requirement.

A Case in Point: Teaching and Learning the Various Forms of the "Rule of Three"

The difficulties facing teachers who wanted their students to develop sound ratio and proportion concepts and skills will be briefly discussed here in order to emphasize the general points made above.

Over many centuries, real-life ratio and proportion problems had been solved by applying the celebrated "rule of three" (see, e.g., Baker, 1568; Recorde, 1543). But, a difficulty facing teachers, and even more so their students, was that the general class of problems that could be solved by the rule of three could be logically categorized into at least four sub-classes. First, there were problems that could be solved by applying the "rule of three direct"; then there were those that needed the "rule of three inverse"; then there were those that required the "double rule of three"; and, finally, there was also a "rule of three composed of five numbers." With each of these sub-classes, there were ways of proceeding depending on whether the context involved whole numbers, common (or vulgar) fractions or, from about 1600, decimal fractions.

Statements of rules for tackling rule-of-three problems were to be found in *abbaci* books, and also in almost every printed arithmetic published during the 17th or 18th century (see, e.g., Baker, 1687; Dilworth, 1797; Fink, 1900; Pike, 1788). Helping students to learn to apply the rule of three, and associated rules, was difficult because the same relationships embodying ratio and proportion concepts arose in many quite different real-life contexts (Dowling, 1829).

Teachers and Students Responding to Difficult Teaching and Learning Environments

Although the best teachers could identify and apply appropriate problem-solving methods to solve each problem, most teachers did not know their subject sufficiently well to do that. In such circumstances, it was easy for cyphering books to become not much more than sources of type problems and exercises with solutions copied from, for example, older cyphering books. There was little opportunity for teachers to devise efficient strategies that would assist their students first to recognize the major category to which a problem belonged, and then to recognize which sub-category, and which sub-sub-category, etc., would be needed.

Students who somehow managed to select and apply suitable methods for solving problems still had to apply skills appropriately in order to obtain correct solutions. They needed to keep track of what they were doing, and when they had finished their calculations they needed to interpret what they had obtained, in relation to the original problems. Students also needed to be able to check to see if their answers made sense in the various problem contexts.

During the 17th and 18th centuries the cyphering-book tradition, which was the basis of the type of mathematics education offered in many private schools and evening classes in England and in Continental Europe, was translated across the Atlantic Ocean into the often struggling communities on the east coast of North America (Carpenter, 1963). There, it would remain alive until well into the 19th century. In the late 1730s and early 1740s, for example, a young boy called George Washington, who was training to become a surveyor but would one day become the first President of the United States of America, prepared several cyphering books (Littlefield, 1904; Monaghan, 2007). Almost a century later, in the early 1820s, another future President, Abraham Lincoln, also prepared a cyphering book (Earle, 1899; Houser, 1943), even though, apparently, he attended school for a total of less than 12 months (Coolidge, 1974). Washington and Lincoln, and others who prepared cyphering books, may or may not have had access to printed arithmetics (Littlefield, 1904).

This translation into the colonies was not without its difficulties. In 1861, Thomas Burrowes, Pennsylvania's Superintendent of Common Schools, stated that "the attempt by one people to adopt the educational system of another, unless the conditions and wants of both be very similar, will either result in entire failure, or delay in success till the imported system shall have assimilated the people to itself" (Burrowes, 1862, p. 250). According to Burrowes, in Pennsylvania, about one-eighth of the common schools were organized in such a way as "to incite the pupil to proper effort" (p. 250). Only one-tenth of the teachers had qualifications that met the legal standard. Many children did not attend school, and of those who did, attendance was often irregular. Common schools were often taught by young women who, in their own school education, had sewn samplers rather than prepared cyphering books (Northend, 1917).

The difficulty of obtaining teachers who were able to teach arithmetic was related to the fact that most members of the community had not studied arithmetic beyond the four operations. In Pennsylvania in 1839, for example, only about 4800 students attended colleges, academies or seminaries, but about 255,000 students attended common schools (Shunk, 1840). In 1840 the Superintendent of Pennsylvania's Common Schools reported that only 44 students in Pennsylvania schools were preparing themselves to teach in common schools (Shunk, 1840). Just how many of these knew anything about arithmetic beyond the four operations on whole numbers was not made clear in the report, but it is likely that the number was very small.

The process by which the cyphering tradition was translated into the North American colonies is discussed in the next chapter. One of the most interesting associated issues is the role that mathematics textbooks played in the various forms of mathematics education that characterized the colonial era. Nathalie Sinclair (2008) commented that "textbooks, especially in the United States, provide a good record of the intended curriculum at any given point in time" (p. 2), but from our perspective, previous historians of mathematics education have tended to concentrate too much on textbooks and textbook authors, and too little on cyphering books and the cyphering tradition. We hope that this book will refocus the history of mathematics education not only on the actual mathematics that students were attempting to learn, but also on what they did in order to learn.

Chapter 3
Translating the Cyphering-Book Tradition to North America

Abstract As soon as formal schooling was established in the North America settlements, the cyphering tradition was adopted as the best means of developing numerical, navigational and surveying skills within the fledgling, largely rural communities. Although, most commonly, boys 10 years of age, or older, prepared cyphering books, some girls also prepared them. The issue of whether the cyphering tradition served the early colonies well is considered—in particular, evaluative functions of cyphering books are outlined.

Educational Realities in the New World

Because of the European background of the settlers, it was inevitable that the cyphering approach to arithmetic education would be adopted in the early European settlements in North America. Despite the inevitability, such an approach was less likely to be as successful in the North American settlements as it had been in Western European nations. In the New World, none of the main requirements for success—necessary *mathematical knowledge* among teachers, sufficient *teaching power*, *student readiness* for learning arithmetic, and adequate *resources*—was usually satisfied. In a particular situation, occasionally one of these four requirements was met, but rarely more than that.

Early North American settlements and states were not ready for structure-based, problem-solving mathematics programs like those at Christ's Hospital, or even like those which had prevailed in the European reckoning schools (Unger, 1888). Most of the settlers had little formal mathematical knowledge, and even into the 19th century relatively few teachers in North America had strong mathematical backgrounds. There was also a lack of teaching power, for many persons who offered tuition in cyphering had had little experience as teachers (Cohen, 2003). Most students were not ready to follow a curriculum inherited from Europe, largely because they had not spent enough time learning elementary mathematics.

Patricia Cline Cohen (1982) has claimed that in the 18th century "local primary schools still did not teach arithmetic as part of their regular course of instruction" (p. 118). However, that may not be completely correct (Small, 1914). Walter Monroe (1917) named quite a few 17th-century local schools that offered cyphering, and it is likely that this practice would have continued into the 18th century. We know, for example, that in 1724 there were four schools in Bruton Parish, Virginia, where girls, as well as boys, learned to read, write and cypher (Perry, 1870, p. 300). Often, boys who started cyphering books were constrained by what they could achieve in the winter months only (Goodrich, 1857). Confined to just four months a year, over a four-year period, boys living in New England colonies or states learned to cypher for no more than 16 months, altogether.

But cyphering skills were obviously useful in colonial society. Indeed, Alice Earle (1899) argued that the attitude toward education of many farmers in colonial North America was that all they wanted their charges to learn from formal education was "the Bible and figgers" (p. 138). According to Cohen (1982), colonial youths were often willing to attend evening cyphering classes, even though they found them boring. Enterprising private tutors took advantage of the personal and potential monetary advantages of being able to cypher.

Money (1993) reported that in 1700 there were at least 17 private tutors within the European settlements in North America who advertised that they were willing to receive students for cyphering, and that between 1720 and 1740 this number grew to 49. Although some of the names of tutors are listed in Louis Karpinski (1980) and Robert Seybolt (1935), the actual number offering classes was likely to have been many times the numbers advertising classes. Indeed, it is likely that there were hundreds of itinerant teachers providing tuition in reading, writing and occasionally cyphering. Some of these itinerant teachers did not read, write, or cypher well, themselves, and they eked out a living as they moved from town to town. The journal of James Guild, "peddler, tinker, schoolmaster, portrait painter," prepared between 1818 and 1824 (Vermont Historical Society, 1937), provides insight into the thinking and activities (and the irregular spelling) of one such an itinerant.

In the 18th century there were far more newspaper advertisements in North American newspapers offering tuition in "arithmetick" (as it was usually written) than advertisements for printed arithmetics (Monaghan, 2007). At that time, arithmetical education was based on students preparing cyphering books. Historians like to mention well-known authors of early European arithmetics (e.g., Recorde, Wingate, Cocker, Hodder, Leybourn, Venema, and Fisher), but it is likely that only a tiny proportion of students who prepared a cyphering book owned, or had ready access to, a printed arithmetic (Goodrich, 1857).

Some cyphering books included pages copied from sections of printed arithmetics, and others may have been copied from manuscripts that were copies of sections of printed arithmetics. But these represented the exception rather than the rule. Many were based on cyphering books created and owned by the master who offered instruction in arithmetic. Students wrote initial solutions to problems on scraps of paper, or slates and if and when the tutor indicated that they were correct, they were inscribed into cyphering books (Monaghan, 2007). Some writing masters made calligraphic headings in their students' cyphering books, and sometimes masters actually wrote exercises to be solved in the books (Parker, 1912). Precisely who made an entry—student or tutor—was not always easy for an outsider to determine.

Robert Seybolt's (1917), Paul Douglas's (1921) and Robert Bremner's (1970) illuminating analyses of apprenticeship education in the North American colonies of the 17th and 18th centuries established beyond doubt that large numbers of male apprentices learned to read, write and cypher in evening classes. Statements in legally enforceable indentures often required masters to ensure that their apprentices learned to read, write and cypher (quotations from indentures are given in Seybolt, 1917). Instruction could be given by masters themselves, but if they were unable to do this then apprentices would be expected to attend day or evening classes.

Most local schoolmasters could cypher, but to varying extents. Often, masters responsible for male apprentices allowed them to attend "school" during certain, "off-seasonal" months only (Bailyn, 1960; Douglas, 1921; Loeper, 1974; Monaghan, 2007; Seybolt, 1917). Sometimes, female apprentices were allowed to attend schools during the day.

Cyphering classes were often held in the late afternoons or evenings, and were attended mainly by apprentices who had risen early in the morning and labored throughout the day. Direct whole-class instruction was a rarity, for most teachers had only a doubtful grasp of the intricacies of the various branches of arithmetic (Parker, 1912). Most students would probably have endured cyphering sessions looking forward to the time when class would be dismissed.

Nevertheless, the cyphering books that were prepared often became important personal reference books. Most apprentices could not afford to buy printed arithmetics, but the cyphering books they prepared became *their* books, their *own* handwritten guides to what they needed to know arithmetically. The apprentices knew that they would be invaluable resources when they completed their indentures. When seeking employment, what better reference could a former apprentice have offered than a neatly and fully developed cyphering book? And many of the cyphering books *did* feature good penmanship and calligraphy. Until the 1830s, most teachers who offered cyphering classes also doubled as writing masters (Thornton, 1996).

Robert Seybolt (1917) emphasized the importance attached to cyphering, by both the masters and the apprentices, in education programs for apprentices. The masters wanted their apprentices to be able to help them with reckoning, measuring, etc., during the day and, for their part, the apprentices wanted to know how to do this so that they would be well qualified when they finished their indentures. According to Seybolt (1917):

> It is interesting to note the content of the course in "cyphering," or arithmetic, pursued by the apprentice. A Westchester indenture of July 1, 1716, makes provision for teaching the apprentice to "read, write & cast accompts to so far as the rule of three." Sometimes this description was added to in the following manner: "Cipher as far as the rule of three direct inclusive." ... In some instances the apprentice was to be taught "to cipher so as to keep his own accounts," or "so far as to be able to keep his booke." (Quoted in Seybolt, 1917, pp. 97–98)

The cyphering books that we have examined (both those in our own principal data set, and those in collections in various libraries) are usually silent on whether they were prepared by apprentices or other young adults attending either evening classes, or by younger students attending common or district schools, or by persons studying privately, or by teachers creating "master cyphering books" for use in their own classes.

It is clear, nevertheless, that throughout the 17th and 18th centuries a large number of boys in the American colonies received instruction according to the cyphering tradition. Although Cohen (1982) asserted that "numeracy did not proceed very far in 17th-century America" (p. 47), and that early colonists were not "a calculating people" (p. 47), Karpinski's (1925) claim that toward the end of the 17th century, and into the 18th century, students in the schools of New England, New York and Pennsylvania learned to cypher and to cast accounts is likely to be closer to the truth. Be that as it may, a claim by David Kamens and Aaron Benavot (1991) that arithmetic became a compulsory subject in primary schools in the United States of America in 1790, and was a compulsory subject in secondary school curricula in the North American colonies as early as 1740, is far-fetched. So too was their claim that the "United States was the first country to establish arithmetic as an important required subject in the public schools" (p. 147).

Cyphering was not confined to New England states, Pennsylvania, or New York. Susie Ames (1958) maintained that in Virginia around 1670, cyphering was part of the curriculum at Syms, Eaton, and other endowed schools, and that some male apprentices in Virginia and the Carolinas learned to cypher from instruction made available to them through the terms of their indentures.

Research by Charles Coon (1915), Louis Karpinski (1980), Robert Seybolt (1917, 1935), E. I. Monaghan (2001, 2007), and Ashley Doar (2006) has demonstrated that throughout the 18th century, North American private schools and private tutors increasingly offered courses in basic arithmetic, mercantile arithmetic, navigation, surveying, and instrumentation. Although some of these classes were held during the day, many of them were held during evenings. In 1725, Rowland Ellis—a Society for the Propagation of the Gospel (SPG) teacher, based in New Jersey—maintained that adults wanted their male offspring, and their male apprentices, to learn to read, write and cypher, in order that they would become well fitted for "trades and employments" (quoted in Monaghan, 2007, p. 152). Ellis reported to the SPG that he had had 17 young male apprentices attending his classes over winter for the purpose of improving their writing and cyphering. Only 21 of the 212 cyphering books in our PDS had a strong emphasis on anything other than arithmetic.

From another perspective, Rebecca Emigh (2002) has argued that all 17th- and 18th-century North American communities engaged in ethnomathematical practices, in the sense that they passed on to children and apprentices the personal numeracies and measurement skills that they would need to fulfil their roles within their well-defined groups. Not everyone could read or write letters and numerals, perform written calculations, or measure in standard ways. However, as in many European nations (Unger, 1888), oral traditions and practical skills acquired in community life taught most people at all levels of society not only to count, but also to use concepts like "more," "less," "the same," etc., to perform actions in appropriate sequences, and to estimate and measure context-relevant distances, times, capacities, areas, and other quantities, when this was required. Harriet Schoen (1938) claimed:

> In the days when the "bounds" of great wilderness tracts were being marked off by deep-cut blazes in the trees along the line, a knowledge of land surveying was a useful skill, and many a boy learned its elements by following the "boundsgoer" in his work of "running the line." And those who did not actually take part in running the line must have attended many a gay springtime "processioning," when neighbours made a festive occasion out of "perambulating the bounds."
>
> Vague land grants and inaccurate surveys made the subject of boundary lines a prime issue in the everyday life of colonial homes, one that caused many arguments and demanded much time and attention. (p. 83)

The question whether it was common for someone to take a formal course in surveying or navigation will be considered later in this present work.

Although standards in navigation or surveying courses offered in North America never matched those at Christ's Hospital, we believe that we have seen the Christ's Hospital influence in several of the cyphering books in our PDS (see, for example, the cyphering books by William Turnbull (1780, Appendix A, m/s #13), Richard Warner (1816, Appendix A, m/s #80), and John Wesley Pierson (1828, Appendix A, m/s #129)). Research by Peter Coldham (1990) has shown that during the period 1617–1778 about 1000 graduates of Christ's Hospital took up apprenticeships in the Caribbean or in North America. Some of them stayed on, directly influencing education development in their new homelands. Cyphering books prepared by students who attended the Royal Mathematical School of Christ's Hospital (London) in the 18th century found their way to North America—we know that at least one such cyphering book is currently held in each of the Phillips Library (Salem), the Clements Library (at the University of Michigan), and the New York Public Library. In our own (Ellerton and Clements') private collection we have two Christ's Hospital manuscripts—prepared by Charles Page in 1825 and Henry Josiah Calkin Andrew in 1836.

Despite the acceptance of the importance of arithmetic education, many leaders within the fledgling North American communities in the 17th century, and especially those with strong religious orientations, thought that, from an education perspective, most children needed only to learn to read the Bible and to remember the catechism (Monaghan, 2007). Boys from well-to-do families were expected to pay attention to the classics, to rhetorical exercises, and to religious knowledge, and the curriculum established at Harvard College reflected that expectation (Cremin, 1970; Monroe, 1922).

In England, the condescending attitude toward mathematics changed, at least at the government level, as the 17th century progressed (Howson, 1982). The last three decades of that century witnessed the emergence of Isaac Newton, Abraham de Moivre, and other talented mathematicians and scientists in England (Taylor, 1954). However, British mathematics still lagged behind mathematics in some of the Continental nations (Craik, 2007), and neither the advances in scientific learning nor the changing attitudes to education of British political leaders were known to many people in the North American settlements (Eby & Arrowood, 1934).

David Eugene Smith and Jekuthiel Ginsburg (1934) wrote scornfully on what, they believed, was the early North American intelligentsia's attitude to all things mathematical:

> The settlers were more concerned with combating the hardships of life in a new country, with religious quarrels, with fighting the bogy of witchcraft, with petty political feuds, with the small gossip of small communities, and with devising tortures (mental and physical), than they were with the masterpieces of contemporary English literature and science. The century that saw the work of Galileo, Kepler, Gilbert, Napier, Fermat, Descartes, Pascal, Huygens, Newton, and Leibniz in countries from which the settlers had come, saw among the intelligentsia [in the North American settlements] no apparent appreciation of the discoveries of scholars of this class. (pp. 14–15)

However, given the rugged pioneering circumstances in the New World (Goodrich, 1857), with would-be scholars completely isolated from developments in European thinking, we believe that it was unrealistic to have expected the situation in the colonies to have been any different from what it was.

Bernard Bailyn (1960) described the number of evening schools in New York in the early 18th century that were offering tuition in cyphering as "remarkable" (p. 33), and added that apprentices were not the only ones to attend these classes. We do not know much about those who taught the classes, but it should not be assumed that all teachers were largely ignorant, and knew nothing about mathematics. We know, for example, that Alexander Malcolm, a renowned British mathematician—whose *New System of Arithmetic,* published in London in 1730, was described by Augustus De Morgan (1847) as one of the most "extensive and erudite" mathematics books of the 18th century—actually operated a private school in New York between 1732 and 1739. In an advertisement in the *New York Gazette* of January 7, 1734, Malcolm declared that he was "Master of the Grammar School" in the City of New York, and taught "mathematicks, geometry, algebra, geography, navigation, and merchants' bookkeeping after the most perfect manner" (the advertisement is reproduced in Karpinksi, 1980, p. 575).

Learning to Count and Measure in the Early European Settlements of North America

Despite the increasing numbers of cyphering classes being offered, especially during winter, in the 17th and 18th centuries, most adults in England and in the distant North American settlements gave little thought to how people should go about learning arithmetic or, for that matter, any branch of mathematics. Most of the children who attended dame schools learned to count and to recognize numerals for numbers from 1 to 10, sometimes with the help of a hornbook (Cremin, 1970; Earle, 1899; Plimpton, 1916; Tuer, 1896). Thus, for example, Stephen Allen, the son of a poverty-stricken widow but someone who would one day rise to become Mayor of New York City, recalled that although, in the 1770s, his early education "was very limited," at the dame schools that he attended he not only learned to read and to write, but also to apply "a few rules in arithmetic" (quoted in Kaestle, 1983, p. 31).

Others learned to count and measure without ever attending school or cyphering classes. Boys working in the fields, and girls working at home, simply learned to perform everyday numerical and measurement tasks (Laughlin, 1845; Rogers, 1990). Although many apprentices, and some slaves, were taught directly to count or to measure by those responsible for their care, often this did not occur (Cohen, 1982; Meyer, 1965; Monaghan, 2007; Morris, 1946).

The extent and quality of arithmetical education varied from place to place, from child to child, from family to family, from school to school, from teacher to teacher, and from apprentice to apprentice. Some cultures valued numerical skills highly. Migrants from Holland and Germany in Pennsylvania, in particular, had a strong reputation for applied number sense (Dewalt, 2006). According to William Parsons (1976), "even those Dutchmen regarded as slow-witted by their neighbours could figure prices and accounts rather accurately," and "by reputation, at least, they rarely accepted a bad bargain" (pp. 120–121).

Although there are no systematic data available on the proportion of children and indentured servants in the 17th-century North American settlements who learned to cypher in or out of school (Meyer, 1965), Monroe's (1917) thoughtful summary of the place of arithmetic in the various systems of education that operated in the different colonies is helpful:

> This survey of the early schools of the American colonies shows that, whether arithmetic was explicitly mentioned along with reading and writing in the official acts of the colonial governments, as in New York, or was omitted, as in the case of Massachusetts and Pennsylvania, arithmetic was taught in the public schools in many towns from the beginning. The activities of trade and commerce, which were centered in these towns, created a demand for arithmetic, and instruction was given in the subject either in public schools or in private institutions. In these schools arithmetic was primarily a tool of commerce. (p. 12)

Monroe (1917), drawing on research by William Eliot Griffis (1909) and William Heard Kilpatrick (1912), pointed out that arithmetic was always included in the curriculum of the Dutch-language schools that were established in New Amsterdam (which, later, would become New York), between the 1630s and the mid-1660s. A parent of a New Netherlands pupil who learned reading paid about 10 cents per week to the teacher, but "writing and cyphering cost about 28 cents a week" (Smith & Ginsburg, 1934, p. 8). In the 17th century, Dutch-background families expected all of their children—male or female—to learn to cypher. Indeed, the Englishman Sir John Child (1693), Governor of the East India Company and President of the Christ's Hospital Board of Governors, maintained that this expectation contributed to what he believed was the superior economic performance of the Dutch nation. Karl Fink (1900) argued that the Dutch were more advanced than anyone else in the arithmetic associated with the exchange of currencies.

According to Monroe (1917), "arithmetic was considered by the early Dutch colonists to be a practical subject, necessary for those engaged in trade and commerce" (p. 6). William Parsons' (1976) and Mark Dewalt's (2006) analyses of the education of children of Dutch and German emigrants in Pennsylvania in the 18th century revealed that most of these children studied arithmetic at school. Thus, although Smith and Ginsburg's (1934) statement that "the facilities for studying arithmetic in and about New Amsterdam were pitifully few" (p. 8) was probably correct, as was their statement that "pupils rarely saw a printed textbook on the subject, their only aids being an occasional hornbook with the numerals" (p. 8), the statements nevertheless give the wrong impression. In fact, one might ask, why should one have expected students to have seen a "printed textbook"? After all, even in Europe it was only in a few prestigious schools (like Christ's Hospital, in London and *l'Ecole Royale d'Hydrographie et Géometrie,* at Le Havre, in France) that school students studied mathematics from a printed mathematics text (Allen, 1970).

The German influence in North America in the 18th and 19th centuries is one of steady growth. The first permanent German settlement was made in Pennsylvania, in the last years of the 17th century. This settlement was followed, in the early 18th century, by others in New Jersey, Maryland and Virginia, and settlements began to extend southwards to the Carolinas and to Georgia, and to New York and into New England. In the 1770s it is estimated that there were about 250,000 German-background people in North America, which was about one-tenth of the total population. A century later, it is estimated that 18 million of the U.S. population were of German-background (Thwing, 1928). These Germans, especially those from Reform-church backgrounds, almost invariably had a strong interest in arithmetic.

Karpinski (1925) pointed out that in colonial North America, cyphering books were prepared by the children of many English and Dutch settlers. These cyphering books were based on "master cyphering books" held by their teachers. However, Smith (1933), in questioning the value of the mathematical and educational outcomes arising from the cyphering tradition, wrote:

> Boys entering apprenticeships in the early days learned enough of calculation to enable them to meet the simple needs of their trade; girls rarely acquired more than the ability to read and write numbers and to perform simple additions. The recruits to the clergy learned a little about the calendar for the purpose of recognizing the holy days; masters of ships learned how to find latitudes by the aid of the astrolabe or quadrant, and to make some crude attempts at finding longitude; physicians often learned how to cast horoscopes and surveyors acquired their trade through apprenticeship or through a meagre training in the elementary schools of Europe. ... Schoolmasters were frequently sent out from the home countries but they were generally a rather worthless lot, able only to teach reading and a little Latin, or to give private instruction in "cyphering" and writing. Outside of the larger towns, pupils seldom saw a printed textbook, and even at Harvard College, which was founded in 1637, mathematics was almost unknown. (pp. 227–228)

That statement is unduly judgmental for there were some remarkable applied mathematical achievements, especially in relation to navigation and surveying, in 17th-century North America (Wroth, 1952). A case in point was Augustine Hermann's map of Virginia and Maryland, first published in London in 1673 (Phillips, 1896). Given the difficult and often harsh circumstances which prevailed in many of the new settlements, and the pioneering and creative spirit needed to achieve anything in education beyond what took place within families in the North American colonies in the 17th century (Meyer, 1965), not much more than was actually achieved might reasonably have been expected.

There are no official population statistics for the North American settlements around 1700, but population estimates for the British colonies at that time vary between half a million and one million (Oswald, 1917). For towns, estimates vary because sometimes slaves were included and often they were not. The three largest towns were Boston (probably about 7000 people), New York (about 5000 people, of whom about half were African-American, for some wealthy families are said to have had as many as 50 slaves), and Philadelphia (about 4500 people, combining emigrants from mainly Sweden and German, often of Quaker persuasion) (Oswald, 1917). Given the circumstances, it is remarkable that in 1700 there were already two colleges—Harvard College, in Cambridge, Massachusetts, and the College of William and Mary, in Williamsburg, Virginia.

From the beginning of the 18th century the population of the North American British colonies steadily increased as did the amount of attention given to formal education. We believe it is important to recognize that the stereotypical disparaging interpretations by Smith (1933) certainly need to be modified for the period 1700–1850.

Thus, for example, not a few 18th-century clergymen were competent mathematicians. Isaac Greenwood, the first Hollis Professor of Mathematics at Harvard College (1727–1737) was a trained clergyman. So too were Dr John Farrar, an early 18th-century Hollis Professor (1807–1836) at Harvard, and Dr Jeremiah Day, Professor of Mathematics and Natural Philosophy at Yale College (1803–1817), President of Yale College (1817–1847), and author of numerous mathematics text books (Ackerberg-Hastings, 2000). It is clear from entries in diaries and cyphering books made by the Reverend Ebenezer Gay (1806, 1812, 1820), a Harvard graduate who became an Episcopal clergyman, that although he loved mathematics, and was good at it, he found it difficult to teach it well to children. The same was true of the Reverend Ichabod Nichols (1784–1859), a tutor in Mathematics at Harvard College in the early 19th century (Nichols, 1921).

It should be recognized that considering the peculiar personal situations in which some early colonial mathematicians found themselves, some of their achievements were nothing short of remarkable. Nathaniel Bowditch, for example, completed a traditional indentured apprenticeship in Salem during the late 1780s and early 1790s. But, what he learned as an apprentice and from a self-imposed, disciplined mathematical reading régime, enabled him to become a world-class mathematician, an excellent teacher of navigation, a brilliant practical navigator, and a pioneering actuary.

The cyphering tradition developed and prospered in North America, despite having been transplanted, by necessity, into rough, remote, and scattered, foreign environs that were very different from those in the European "homelands." Even at the time of the first national census, in 1790, the three largest cities in the United States—New York, Philadelphia, and Boston—had populations of about only 33,000, 29,000 and 18,000, respectively—although, not all inhabitants were counted for census purposes.

Making connections between education establishments in different population centers was difficult. Around 1700, for example, it took about four days to travel from Boston to New York or from New York to Philadelphia (Oswald, 1917). Given the circumstances, the creation of a sound system of mathematics education in the colonies, a system whose influence extended beyond one town or region, was an extremely difficult thing to do. Of necessity, schools became integral parts of the cultural artifacts of the early communities. One of the aims of this book is to recognize and celebrate the educational, academic and practical achievements of those who helped establish the cyphering tradition in North America. Another aim is to draw attention to educational strengths and weaknesses of practices associated with that tradition.

Why Study Arithmetic Beyond the Four Operations on Whole Numbers?

Why might some colonial adults want their offspring, or the apprentices in their charge, to study arithmetic beyond the four operations on counting numbers? One simple answer is that there were many everyday tasks in which a working knowledge of arithmetic up to, and beyond, the various rules of three could be profitable. Consider, for example, the arithmetic that might be needed to operate the *Adventure*, a sloop owned by Christopher and George Champlin, wealthy traders living in Newport, Rhode Island.

The Champlins engaged in what was known as the "triangular trade," and the *Adventure* made four transatlantic trips during the first half of the 1770s (Rawley, 1981). The Champlins and their partners would send the *Adventure* laden with rum and manufactured goods supplied by New England merchants. On the first trip, in 1770, for example, the crew comprised the captain, first and second mates, a navigator and seven sailors (Rawley, 1981). In Africa, the cargo was bartered for slaves (for example, about 200 gallons of rum was exchanged for a male slave, and about 150 gallons for a female slave), Seventy slaves were obtained. Then, the sloop proceeded to the West Indies (Barbados, Grenada, and Jamaica), where the 67 slaves who remained alive were sold for between 35 and 40 pounds a head, on average. Molasses was then purchased, and the final leg was back to Newport, where the molasses was made into rum, and sold—locally, or on the next voyage to Africa. James Rawley (1981), when commenting on the third of the *Adventure's* trips, stated:

> In assessing the profitability of this voyage, it is to be noted that customarily the entire outfit and one-third of the original cost of the vessel were a debit. Other costs included wages, commissions paid to the captain and agents in Grenada, expenses on the African coast, and insurance. The vessel brought back molasses and nine sets of bills ... The estimated profit on this venture, which had lasted a year, was 400 pounds, or about 23 per cent. (p. 372)

Much careful preparation for a journey was needed, and in particular many arithmetical calculations needed to be made and recorded. The sloop and its cargo needed to be insured, wages negotiated for crew, agreements reached among participating entrepreneurs on how profits, or losses, would be shared, and cargo prepared and placed on the sloop. Often, between 15,000 and 30,000 gallons of rum needed to be committed to barrels. James De Wolf, a Boston and Rhode Island-based slave trader, operated a distillery that, on average, turned 300 gallons of molasses into 250 gallons of rum, every day. This was stored in hogsheads—largish wooden casks, the extent of which became the name of a unit of capacity. Sometimes, rum was "watered down," and calculations had to be made using a branch of arithmetic known as "alligation" (an old and practical method of solving arithmetic problems related to mixing ingredients).

Rum was not the only product exported from the colonies. Horses, spermaceti candles, and fish, were also important, and served as exchange for European necessities and fineries not readily available in the New World (Andrews, 1912; Rawley, 1981). According to a Peabody Essex Museum (undated) publication, Salem merchants traded lumber, hides, masts, wool, cod, and their best rum for salt, linen, hardware, and bar iron (p. 20). The planters in the South sold tobacco to the world. Between 1703 and 1788, for example, freight receipts for the American colonies on tobacco averaged 82,000 pounds sterling per annum (Rawley, 1981).

The Champlin brothers, and other venturers, sought to maximize profits from their risky trade activities by employing intelligent, arithmetically-competent persons and basing them both in company offices and on ships, to oversee likely costs and risks. These "on-board" specialist arithmeticians, known as a "supercargoes," not only needed to know details about restrictions on trade incorporated in navigation acts (Andrews, 1912), but also had the vital responsibility of bartering for profit whenever trading took place during journeys (Bartlett, 1832; Gibbs, 1931). Over a 12-month period in 1796–1797, for example, a young Nathaniel Bowditch was supercargo on a venture to the East Indies, and he made extensive notes both on the calculations he needed to do, and on the contexts in which these were carried out (Bowditch, 1797).

For international shipping ventures there were many hidden costs: for a triangular trade episode, for example, guns might need to have been fitted, handcuffs and shackles purchased or repaired, large amounts of provisions purchased and packed, and much vinegar, needed to clean slave quarters, placed in containers. Furthermore, a skilled navigator, capable of guiding a brig safely to its destination in the fastest time possible, was needed. Competent and experienced navigators could be hard to find, for navigational accuracy required considerable arithmetical and geometrical expertise, as well as knowledge of instruments and their operations.

The Practical Emphases in Colonial Cyphering Books

In colonial North America there were many aspects of everyday life in which well-developed arithmetical and geometrical skills were useful. Land needed to be surveyed, buildings designed and constructed, loans negotiated, and accounts established and maintained. With so much family and village commercial activity relying on arithmetic or geometry, and transportation between the New World and Europe and Africa so heavily dependent on shipping and navigation, one would have expected families to encourage young boys to learn business-related arithmetic and geometry.

But what about the girls? Most (although, as we shall see, later, not all) colonial families thought that girls did not need to learn much about numbers beyond being able to count orally and, perhaps, to add and subtract. A family struggling to survive financially could be excused for thinking that the more time children spent at school the less time they spent on tasks that would help put food on the table. As a result, in many New England families boys were sent to schools in winter months, only. At school, time spent on preparing cyphering books could only be justified if the arithmetic content was "practical." For girls, time spent at school on sewing samplers was usually thought to be more easily justified than time spent on cyphering, because sewing was important in families. Only a small proportion of families—most, but not all of whom had Continental European backgrounds—wanted girls to do arithmetic beyond numeration and the four operations on whole numbers (Smith, 2005).

The popularity of annual almanacs testified to the advantage that arithmetically literate people had in meeting the daily challenge of coping with the numerical challenges of everyday life. Dudley Leavitt (1772–1851) a teacher of mathematics but, more famously, a writer and publisher of almanacs, had a strong arithmetical emphasis in his *Farmer's Almanacks* which were published every year (and sometimes twice yearly) between 1797 and 1851. In the 1840s, 60,000 of his almanacs were sold each year. In his *New-England Farmer's Almanack and Agricultural Register for the Year of our Lord 1820* (Leavitt, 1820), for example, there were many tables showing numerical data relevant to the everyday lives of readers.

One such table showed the value of a Federal dollar in different state currencies—for the states had different currencies. Another table showed daily interest, at 6 percent per annum, on loans from 1 to 12,000 dollars; another showed the "values of gold coins of Great Britain, France and Spain according to the Act of Congress, April 29, 1816." It is simplistic to criticize the arithmetic found in cyphering books as irrelevant to the lives of ordinary people. To survive with dignity, adults who were engaged in agriculture or business pursuits needed to be numerate. They needed to be able to use numbers themselves, and to interpret charts of numbers found in almanacs.

The mercantile spirit of the day was often captured in model problems and exercises in students' cyphering books. Our principal data set (PDS) for the main analysis carried out later in this book includes perhaps the two earliest extant cyphering books prepared in British colonial North America, and it will be useful, here, to consider briefly the kind of arithmetic to be found in those two books, as well as in some later cyphering books.

The Chichester-Pine family cyphering book. The oldest, clearly dated, cyphering manuscript in our PDS, and probably the oldest extant cyphering book prepared in North America, is part of a composite cyphering book with about 300 handwritten pages, the first nine of which were written in 1701 (Appendix A, m/s #1). The separate sections that formed the composite manuscript were prepared, at different

Why Study Arithmetic Beyond the Four Operations on Whole Numbers? 49

times, by members of the Chichester or the Pine families. The Pine family, originally from Connecticut, was one of the earliest families to settle in Huntington, Long Island. That was where this cyphering book was prepared.

The precise time (or times) when the different sections were sewn together is unknown, so there can be no guarantee that the content as it is now presented in the composite manuscript is in the chronological order in which it was written. The first section, at the beginning of the manuscript, is clearly dated 1701, but we think that an undated 54-page section, also sewn into the manuscript, could have been prepared in the 17th century. Most of the content in the 312 pages comprises rules, cases and problems directly related to the practical needs of families and to business concerns in the North American colonies. The headings in the 1701 section are addition of money, avoirdupois weight, troy weight, addition of time, of liquid measure ("liquid measure is of two sorts, viz: one for wine, brandy, &c, and the other for beer and ale"), dry measure, long measure, land measure, and compound subtraction.

The second of the sections sewn into the Chichester-Pine manuscript is undated but was probably prepared some time during the first half of the 18th century. It features particularly high quality penmanship and calligraphy, and the arithmetic is more advanced than that in the first section. Topics covered include compound multiplication, the rule of three direct, practice, simple and compound interest, fellowship, compound fellowship, and loss and gain.

"Fellowship" was concerned with family and business partnerships in which different persons contributed different amounts of capital, for different periods of time. In the Chichester-Pine manuscript the concept of fellowship was introduced through the following rhyme:

> As the sum of the several stocks
> Is to the total gain or loss,
> So is each man's share in stock
> To his share of the gain or loss.

A similar rhyme was given at the beginning of the section on compound fellowship. A typical compound fellowship problem solved in the Chichester-Pine family manuscript was:

> Three merchants traded together. A put in 120 pounds for 9 months; B 100 pounds for 16 months; and C 100 pounds for 14 months. They gained 100 pounds. How must it be divided?

The statement of the problem in the manuscript, associated calculations, and a statement of the answer, are shown in Figure 3.1.

Figure 3.1 illustrates the problem-calculation-answer genre (hereafter "PCA"). A prevailing feeling was that arithmetic texts were most useful if they did not provide "unnecessary" details. Charles Marshall (1789) explained that "a definition and a rule" should be expressed "as plain, concise and general" as possible, "so that there is no more than what a boy may conveniently copy into his cyphering book" (p. v). With PCA genre, a concise written statement of a problem was to be followed by calculations and a succinct statement of the answer. No explanations were expected. In Figure 3.1, the answer is given immediately after the problem statement. This genre was characteristic of almost all 18th- and 19th-century cyphering manuscripts.

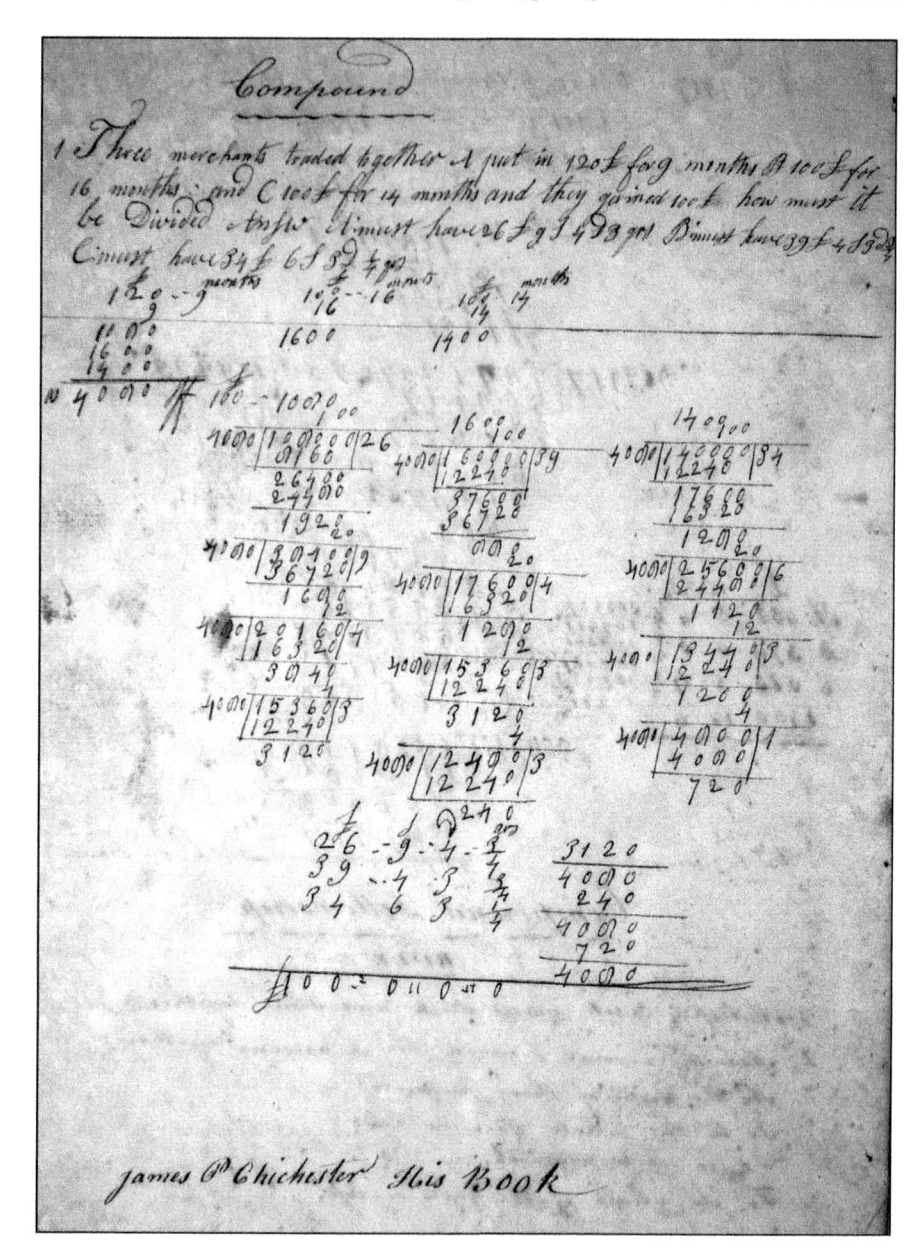

Figure 3.1. A compound fellowship problem from an early North American cyphering book.

The third section sewn into the Chichester-Pine manuscript is undated, but, as we stated above, we believe that the calligraphy and penmanship are more typical of the late 17th than of the early 18th century. Part of a page from a section on "reduction" is shown in Figure 3.2.

Why Study Arithmetic Beyond the Four Operations on Whole Numbers? 51

Reduction

4 farthings makes 1 peny
12 pence 1 Shilling & Qo...
Shillings makes 1 pound

In 1000 pounds. of Demand
how many Shillings pence and farthings

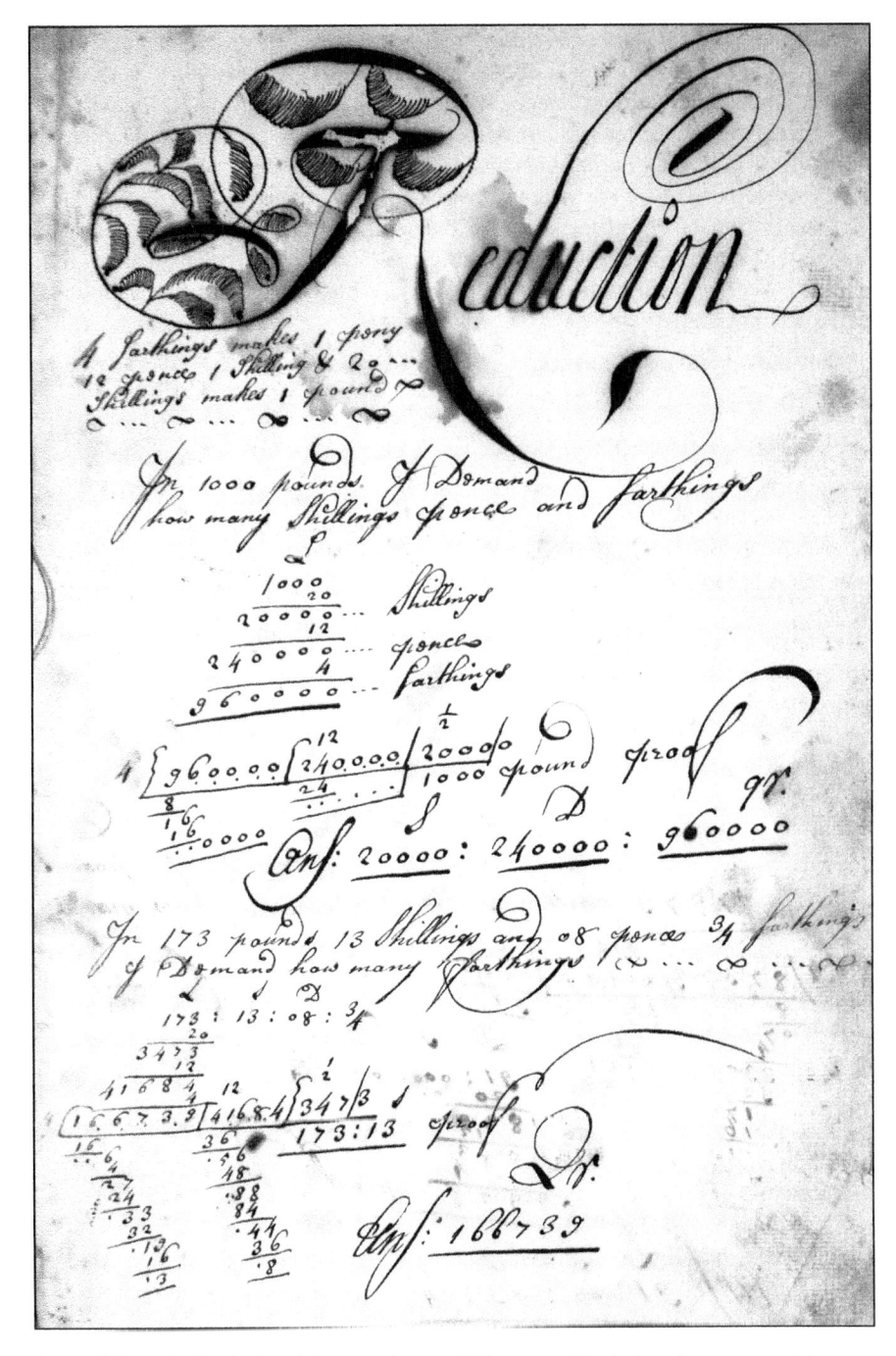

Figure 3.2. A "reduction" task from, perhaps, a 17th-century North American manuscript.

Following the statement of the rule: "4 farthings make 1 penny, 12 pennies make 1 shilling, and 20 shillings makes 1 pound," an exercise is given: "In 1000 pounds I demand how many shillings, pence and farthings." A PCA genre is evident, with the answer being stated after calculations. The word "proof" also appeared after the calculations. Following European practice (Fink, 1900), whenever "proof" was used in a cyphering book it simply meant "a check of the working."

Another problem from the undated, but perhaps 17th-century, section in the Chichester-Pine manuscript was structurally similar to a type of mathematics problem that, over the centuries, became well known. With the original spelling retained, the problem was:

> Admitt a hogshead hath 3 cocks, *A, B, C,* to evacuate water or any other liquor. Lett the Cock *A* alone run, the hogshead will be empyted in 60 hours. With Cock B alone in 30 hours; Cock *C* alone in 15 hours. I demand in how long time the hogshead will be emptying all 3 cocks, *A, B, C* running together.

The solution to this problem occupied a large page (dimensions 13″ by 8″). PCA genre was evident, with many calculations but no explanations. The answer obtained was "8: 34: 1/4: 1/7," which, presumably, meant a little more than 8 hours and 34 minutes. A "proof" was shown.

Thomas Prust's (1702) cyphering book. The second-oldest cyphering book in our PDS is clearly dated 1702 and was prepared by a certain Thomas Prust (who sometimes wrote his family name as "Priest"). Just where Prust lived was never specifically stated, but Internet research has revealed that he probably prepared his manuscript somewhere in Maryland, where the Prust family name was common around 1700. The following problem is taken from a section in his manuscript on "addition"—once again, the original spelling is retained:

> A merchant hath 4 ships safely arrive home this year 1702 in which he had several goods sent in the good ship called the *Endever* which came from Lisborn, 80 chests of sugar, and 250 tapnals of figs; and in the good ship called the *Exchang*, which came from Lisborn, 52 chests of sugar and 125 tapnals of figs; and in the ship called the *Prosperous*, which came from Lisborn, 564 bars of iron and 30 bags of wool, and in the *Green Dragon*, 431 bars of iron and 22 bags of wool. Now the question is to know how many chests of sugar, tapnals of figs, bars of iron, and bags of wool this merchant hath had come home severally in all these 4 ships. (p. 2, Thomas Prust's (1702) cyphering book, Appendix A, m/s #2)

In the 17th and 18th centuries, North American merchants consistently brought goods to the American colonies from Lisbon (Portugal) (Lounsbury, 1930; Lydon, 1965), and on page 7 of Prust's (1702) manuscript there is a reference to the *Maligoe*, a ship which traded between Europe and the American colonies.

There are other word problems in the Prust manuscript that refer to fish and flour having been sent to Europe from Newfoundland, a practice captured in the title of James Lydon's (1965) article, "Fish and flour for gold: Southern Europe and the colonial America balance of payments," and emphasized by John McCusker and Russell Menard (1985) in *The Economy of British America, 1607–1789*. A 1745 article, titled "A General View of the Conduct of the French in America and of Our

Settlements There," which appeared in volume 25 of the London-based *Gentleman's Magazine*, emphasized the importance to the British economy of the American colonies trading fish to European cities in exchange for products like wool.

Nearly all of the problems in Prust's manuscript were related to everyday events. In a section on "multiplycation practice," for example, the following delightfully-worded problem was solved (original spelling retained):

> A tayler mared his daughter to a sayler and gave her 100 needels and to every needle he houngs 20 threads and to every end of every threed houngs 12 purses and in every purs he puts 12 penses. Now the question is, what portion of money the tayler gave with his daughter in marage to the sayler?

Figure 3.3 shows Thomas Prust's (Appendix A, m/s #2) statement of the problem and his solution. PCA genre is evident.

Thomas initially multiplied 100 by 20 using a long multiplication algorithm, and then he multiplied this result by 2 (because each thread had two ends). He then multiplied by 12, and again by 12 (each time using the long multiplication algorithm), to arrive at 576,000. This gave the number of pence, which was then converted to 48,000 shillings by dividing by 12, and then to 2400 pounds by dividing by 20. The algorithm for division is difficult to untangle, but it is one that Thomas often used in his cyphering book.

Content in other cyphering books. Sally Halsey's (1767) cyphering book (Appendix A, m/s #4) is interesting because she dealt with problems that were clearly beyond the elementary numeration and four-operation tasks that normally marked the upper limit of arithmetic for girls. Her book contained numerous

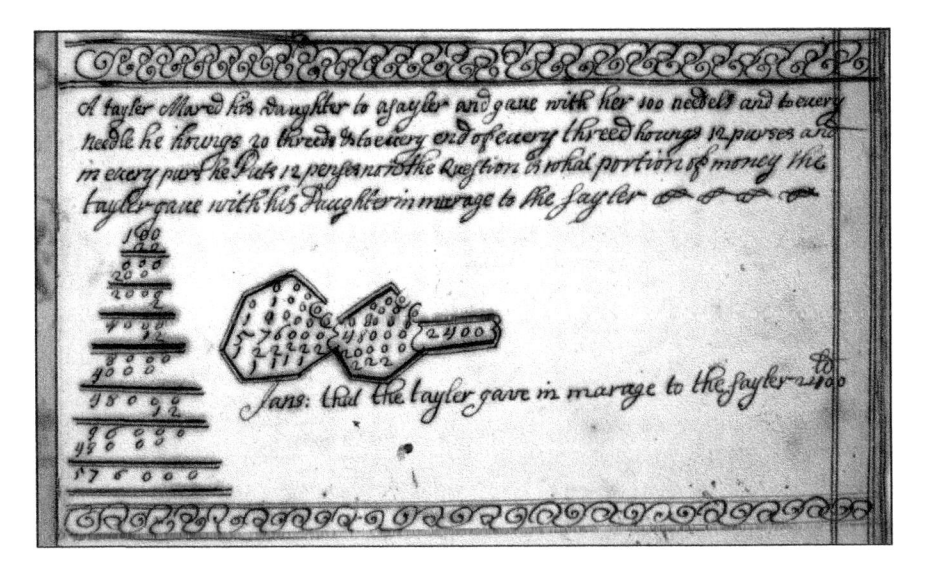

Figure 3.3. An example of PCA genre from Thomas Prust's (1702) cyphering book.

examples that were linked to everyday life, and to business. Thus, for example, Sally considered the following question:

> If, when the price of a bushel of wheat is 6 shillings and 3 pence, the penny-loaf will weigh 9 oz, what must the penny-loaf weigh when wheat is 4 shillings and 6 pence per bushel? (p. 25)

She answered the following question in a section headed "compound interest":

> What is the compound interest of 400 pounds 10 shillings at $3\frac{1}{2}$ per cent per annum for three years? (p. 57)

She also dealt with the following "equation of payments" task:

> D owes E 100 pounds, whereby 50 pounds is to be paid at 2 months and 50 pounds is to be paid at 4 months. But, they agree to reduce them to one payment. When will the whole be paid? (p. 67)

The following appeared in Sally's book, in a section headed "alligation total":

> A vintner hath several sorts of wine, viz Canary at 10 shillings per gallon, Malaga at 8 shillings, Rhenish at 6 shillings and Ohorto at 4 shillings. He is minded to make a composition of 60 gallons worth 9 shillings per gallon. I demand how much of each sort he must have. (p. 89)

David Townsend, of Long Island, New York, solved an almost identical alligation problem in his 1770–1774 cyphering book (Appendix A, m/s #5)—the only difference between the wording of the two problems being that with David's problem the composition needed to be worth 5 shillings per gallon. Another alligation problem recorded by David was concerned with mixing teas:

> If I have 4 sorts of tea—one at 9 pence per oz., another at 12 pence, another at 24 pence and another at 30 pence—how much of each sort must I take to make the compound worth 20 pence per oz.?

The solution to this problem, as it appeared in David Townsend's cyphering book, is shown in Figure 3.4.

The numeral 20 on the left of Figure 3.4 was the given price per ounce of the compound; the next column shows the given prices per ounce of the four kinds of tea; the third column shows the results, listed in reverse order, obtained by subtracting 20 from each of the two larger numbers in the second column, or subtracting each of the two smaller numbers in the second column from 20 (i.e., $30 - 20 = 10$, $24 - 20 = 4$; $20 - 12 = 8$, and $20 - 9 = 11$). These entries in the third column, 10, 4, 8, 11 defined the required ratio, 10: 4: 8: 11 (shown in the fourth column). The calculations on the right represented a check. The procedure was based on an algorithm that can be easily justified by algebra, but it is unlikely that David Townsend would have wondered where the procedure came from, why it worked, or whether the solution was the only one possible. His main task for the future was to remember the algorithm and to recognize the structural features of alligation tasks which called for that particular algorithm to be applied.

Why Study Arithmetic Beyond the Four Operations on Whole Numbers? 55

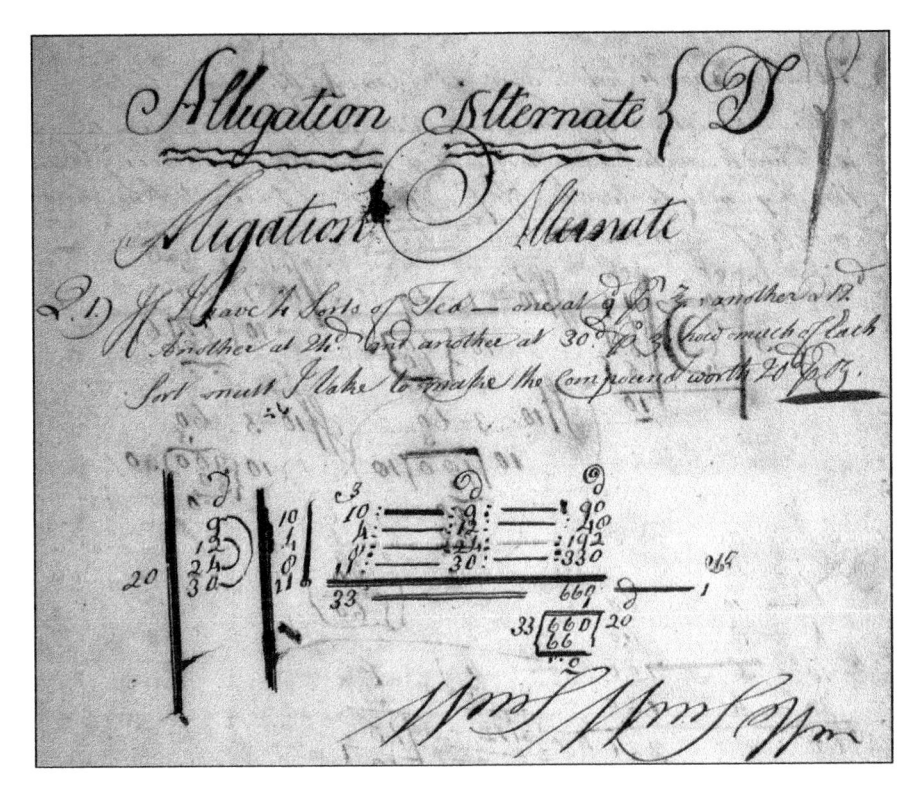

Figure 3.4. Solution to an alligation word problem by David Townsend (1770–1774).

Olney Brayton's (1792–1794) 90-page cyphering book was prepared in Rhode Island. From a content perspective, the arithmetic dealt with themes that were relevant in Rhode Island's mercantile environment. It began with the four operations on whole numbers and quickly moved through to the rules of three—direct, inverse and double—and then to topics like reduction, measurement, fellowship, equation of payments, and barter. The "golden rule" (or "rule of three"), which defined the standard approach to the crucially important study of proportionality throughout the period 1200–1850, was introduced through the following poem:

The golden rule hath places three,
The first and third must so agree,
That of one kind they may remain,
If to the truth thou will attain.
Then third by second multiply,
Divide by the first ingeniously,
Then will your quotient shew the same,
Which you in proper place must frame.

Immediately after this was the problem: "If 1 yard of checks cost 2 shillings and 3 pence, what will 50 yards cost at that rate?" The solution was generated as follows:

If 1 ____ 2s 3 d _____ (yds)
         12
         27
         50
    12 1350   112   5   Answer 5 pounds, 12 shillings 6 pence
         12   100
         15   12
         12
         30
         24
          6

No explanation was given—the figures were written down in the manner shown. Following this model example, there were 11 pages of progressively more difficult single rule-of-three exercises, before the "rule of three inverse" and the "double rule of three" were introduced.

When a new topic was being introduced in a cyphering book, an introductory statement would usually be given. Thus, for example, when the topic "tare and tret" was broached for the first time, Thomas Burlingame Junior, of Providence, Rhode Island (Appendix A, m/s #11) wrote the following explanation (see Figure 3.5, in which original spelling has been retained) in his cyphering book, which was prepared between 1779 and 1784.

Figure 3.5. Tare and trett—in Thomas Burlingame Junior's (1779–1784) cyphering book.

Why Study Arithmetic Beyond the Four Operations on Whole Numbers? 57

The introductory statement shown in Figure 3.5 stated:

> Gross weight is the weight of the goods in hundreds, quarters and pounds with the weight of the hogshead, cask, chest, bag, bale, &c that contained the goods. Tare is allowed to the byer for the weight of the hogshead, cask, chest, bag, bale, &c. trett is an allowance made for most dust in sundry sorts of goods such as tobacco, cotton, pepper, spices, &c, and is always 4 lbs per 104 lbs suttle and found by dividing the suttle pounds by 26 because 4 times 26 makes 104. When the gross weight is brought into pounds and before the tare is deducted they are called pounds gross and after the tare is subtracted the remaining pounds are called pounds suttle, which divided by 26 as said before quotes pounds trett, the allowances for tare are variously wrought.

Immediately after this statement came a model example: "In 12 casks of indigo containing 45 c. 1 qt.14 lb gross tare 30 lb per cask. How many pounds neat?"

In Thomas Burlingame's solution, the 45 hundredweight and 1 quarter were converted to 181 quarters, and these were multiplied by 28. With the 14 lb added, this came to 5082 pounds, and from this 30 times 12 pounds, or 360 pounds, were subtracted (as tare), to arrive at the answer "4722 lb neat."

The wording of the introductory statement on tare and trett was complex. But boys destined to become reckoners on the wharves in Newport, or Boston, or Salem, or Newburyport, or Philadelphia, or New York, or Baltimore, etc., got to know that terminology quickly. That was especially true of those who worked in the customs houses during non-winter months.

Laws of single and double position appeared in some cyphering books. The manuscript prepared by 14-year-old Seth Green Torrey (1788) (Appendix A, m/s #18), for example, recorded the law of double position as: "If the error be alike, i.e., both be too small or both be too long, divide the difference of the product by the difference of the error and the quotient will be the answer. If the error be unlike, that is, one too small and the other too large, divide the sum of the product by the sum of the error, and the quotient will be the answer."

Although it would have been impossible for most early American farmers to comprehend why such a law worked, its relevance in a farming community was suggested in the following model example that appeared in Seth Green Torrey's (1788) manuscript.

> A certain man having drove his swine to market (viz) hogs, sows and pigs, received for them all 50 pounds, being paid for every hog 18 shillings, for every sow 16 shillings, and for every pig 2 shillings. There were as many hogs as sows and for every sow, 3 pigs. I demand how many there are of each sort.

The solution shown in the manuscript started by assuming that there were 6 hogs. The cost of 6 hogs, 6 sows and 18 pigs would have been 12 pounds, which was a "negative error" of 38 pounds. The second "position" was to assume that there were 10 hogs. The cost of 10 hogs, 10 sows and 30 pigs would have been 20 pounds, or a negative error of 30 pounds. The difference of the products (380 − 180 = 200) divided by the difference of the errors (38 − 30 = 8) would give 25. Therefore, the correct number of hogs was 25, and it would follow that the number of sows was 25, and pigs 75. Seth Green Torrey's (1788) handwritten solution is shown in Figure 3.6 (in which he did not state, explicitly, that his second assumed position was 10 hogs).

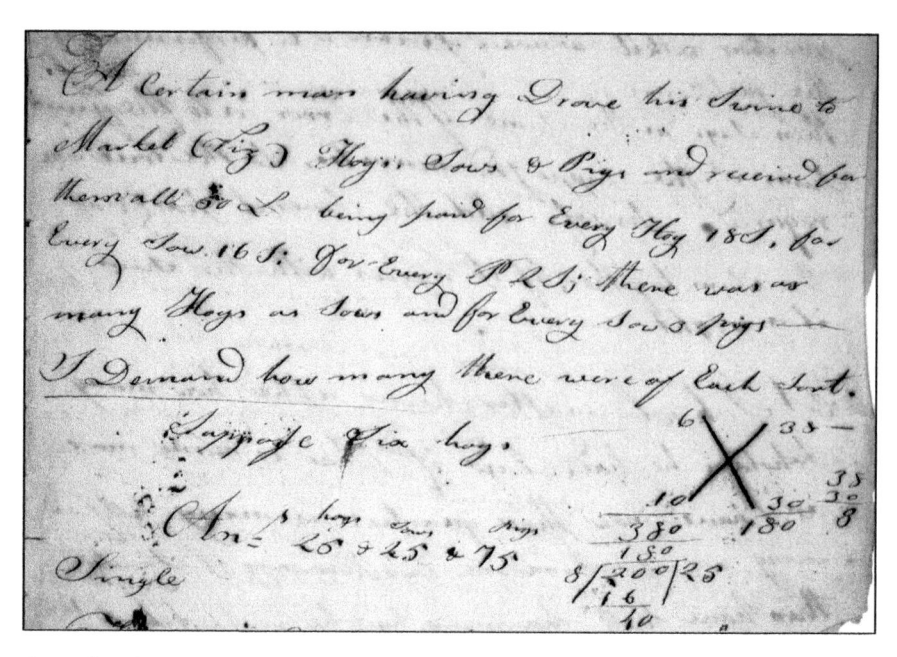

Figure 3.6. Solution to a "double position" problem—Seth Green Torrey's (1788) manuscript.

Why such a method worked, and the structure of problems for which it worked, did not seem to be of much direct concern to anyone. This was a standard type of problem, and students were expected to learn how to apply the method by observing a model (or "type") example, and then, as exercises, solve numerous similar problems passed on from antiquity.

Another topic, known as "gauging," was concerned with finding the capacity, usually in gallons, of a wooden cask. A rule for doing this, according to Daniel Adams (1822), was:

1. Add two-thirds of the differences between the head and bung diameters to the head diameter for the mean diameter; but if the staves be but little curving from the head of the bung, add only six-tenths of this difference.
2. Square the mean diameter, which multiplied by the length of the cask, and the product divided by 294, for wine, or by 359 for ale, the quotient will be the answer in gallons (p. 203).

Nicolas Pike (1788) elaborated upon the "two-thirds of the differences" rule mentioned in the first of these statements by commenting: "If the staves of the cask be much curved or bulging between the bung and the head, multiply the difference by .7; if not quite to curve, by .65; if they bulge yet less by .6; and if they are almost, or quite straight, by .55, and add the product to the head diameter; the sum will be the mean diameter" (p. 459).

A typical example might have been to calculate how many ale- or beer-gallons a cask would hold if it had a bung diameter of 31 inches, a head diameter 25 inches, and a length of 36 inches. Where the standard rules for gauging came from was not discussed in textbooks on arithmetic, and so a student either had to remember the rule or find it in a book. Most commonly, that book would have been the person's own cyphering book. Certainly, sections on gauging can be found in many cyphering books prepared in mercantile centers—of the cyphering books in the PDS, for example, those by Leonard Levi (1810, Appendix A, m/s #57), Richard Warner (1816, Appendix A, m/s #80) and John Wesley Pierson (1828, Appendix A, m/s #129), prepared in Boston, Pennsylvania, and Maryland, respectively—dealt with gauging.

In the preceding discussion special attention was given to colonial arithmetical topics—such as fellowship, the various rules of three, gauging, tare and tret, single and double position, and alligation—that, in the 21st century, are no longer part of school mathematics curricula. What makes those topics especially interesting, from a historical perspective, is that they were all part of the *Liber Abbaci/abbaco* curriculum tradition which had originally been developed within Hindu/Arabic nations and passed on to generations of European merchants, reckoners and children through the cyphering tradition (Swetz, 2005; Van Egmond, 1976). That tradition was translated, almost intact, to the American colonies. But, as we shall see, there would come a time when that tradition would be challenged, and its demise would come quickly.

Other topics, also part of the standard *abbaco* curriculum, were also found in North American cyphering books throughout the 18th century and up to about 1860. Almost all colonial cyphering books that focused on elementary arithmetic included entries on compound operations, currency exchange, reduction, vulgar and decimal fractions, and percentage. If more advanced topics were included, then questions on loss and gain, barter, brokerage, annuities, duodecimals, equation of payments, mensuration, arithmetical and geometrical progressions, permutations and combinations, and involution and evolution were likely to be found. With the exception of decimal fractions and percentage, whose popularity grew both in Europe and America from about 1650 onwards (Whitrow, 1988), all of these slightly more advanced topics had always been part of the *abbaco* tradition. Some writers (e.g., Cajori, 1890; Doar, 2006) have asserted that the mercantile emphasis in the early North American arithmetic curriculum was unfortunate, and arose because "the sins of early English pedagogues" were "visited upon the children in England and America" (Cajori, 1890, p. 50). Such a statement failed to take into account the Continental origins of the *abbaco* curriculum.

Most teachers or private tutors who taught cyphering in the 18th century in the New World accepted this business-related curriculum as standard and necessary. Thus, for example, in the early 1790s, "sundry teachers in and near Philadelphia" prepared *The Tutor's Assistant, or a Compendious System of Practical Arithmetic* (Todd, Jess, Waring, & Paul, 1794), a text that would go through many editions (until 1830) (Karpinski, 1980). Its content was consistent with *abbaco* traditions, and the practical inclinations of the authors were revealed when they included 31 pages of appendices on book-keeping. The title page for this book proclaimed that the book was prepared for children in American schools, suggesting that the *abbaco* tradition had infiltrated common schools.

Michael Walsh (1820), whose *New System of Mercantile Arithmetic adapted to the Commerce of the United States* would appear in many editions between 1801 and 1832 (Karpinski, 1980), was among many other popular arithmetic authors whose texts had a strong *abbaco* flavor. Indeed, the *abbaco* tradition could be found in all parts of English-speaking North America. Thus, for example, the expanded title of an arithmetic text authored by Elijah H. Hendrick (1810) and published in Richmond, Virginia, was: *A new and plain system of arithmetic, containing the several rules of that useful science, concisely defined and greatly simplified. The whole, particularly adapted to the easy and regular instruction of youth and to the trade and commerce of the United States.*

In the above discussion, little mention has been made of geometry and trigonometry, but some of the manuscripts in our PDS dealt specifically with those aspects of geometry and trigonometry that related to navigation and surveying. The largest North American collection of cyphering books relating to navigation education is to be found at the Phillips Library in the Peabody Essex Museum, Salem, Massachusetts. Most of the manuscripts in that collection were prepared in and around Salem itself (Gaydos & Kampas, 2010). In the next section we take a close look at that collection, and especially at those cyphering books in which the focus was on any or all of geometry, trigonometry and navigation.

Practical Emphases in Arithmetic and Navigation Cyphering Books

Robert Middlekauff's (1963), *Ancients and Axioms: Secondary Education in 18th Century New England* included the following statement regarding cyphering books (which he termed "copybooks"):

> Copybooks reveal much about curriculum and teaching. The libraries of Yale and Brown have a few, as do the Massachusetts, Connecticut and Rhode Island Historical Societies. Others may be seen at the Forbes Library and the New Haven Colony Historical Society. But the best collection, numbering several hundred navigation, surveying, arithmetic, and mathematical copybooks, is in the uncatalogued MSS in the Essex Institute. (p. 197)

After having examined extant collections of cyphering books in public libraries and archives across the nation we would agree with Middlekauff's observation that the cyphering book collection held by the Phillips Library (formerly Essex Institute Library) is the largest and best held in a public institution in the United States. During a period of two weeks at Phillips Library in 2009 we were able to locate 110 manuscripts that satisfied our definition of a "cyphering book" (Ellerton & Clements, 2009a). Subsequent searching by research officers at Phillips Library added significantly to that number, so that 201 books, of which 195 are cyphering books, have now been located (Gaydos & Kampas, 2010), and more are likely to be found.

Of the 195 cyphering books located in the Phillips Library in 2011, 130 were totally, or primarily, concerned with arithmetic, and of those, 95 were originally written in or around Salem. Of the other 35 cyphering books concerned with arithmetic, 14 were prepared in other locations in Massachusetts, 14 in New Hampshire, 3 in Maine, 4 in Rhode Island, 1 in Canada, and 1 in Great Britain (Gaydos & Kampas, 2010). Thus, the Phillips Library collection of arithmetic cyphering books comprises manuscripts that were mostly prepared in New England. Of the 130 cyphering books dealing with arithmetic, 29 were definitely, or probably, prepared before 1800.

The Phillips Library collection of cyphering books that focus on geometry, trigonometry or navigation comprises 65 manuscripts, most of which were prepared in or around Salem. That is easily the largest collection of extant geometry, trigonometry or navigation cyphering books in the nation. Most of these books start with elementary Euclidean geometry (definitions and constructions), proceed to trigonometry, and then conclude with a large section on navigation. Some of them are beautifully illustrated in ink, and some illustrations are presented as watercolor paintings (Gaydos & Kampas, 2010).

Practical Emphases in the Phillips Library Cyphering Books

Historic Salem. In 1790, the thriving Massachusetts seaport of Salem, with a population of 8000, was the sixth largest, and the richest per capita, city in the United States of America (Flibbert, Goss, McAlister, Tolles, & Trask, 1999; Peabody Essex Museum, n.d.). Partly as a result of its clandestine but nevertheless active participation in the transatlantic slave trade (Rawley, 1981), and partly through creative development of extensive trading links with Europe, China, India, Sumatra, the Philippines, Canada, and other U.S. ports, Salem had became a major mercantile center that needed a continual supply of competent reckoners, coopers, and other skilled tradesmen to service its waterfront operations (Bean, 2001; Nichols, 1921). It also needed a steady supply of well-trained navigators for its sloops.

America's greatest-ever navigator, Nathaniel Bowditch, acquired his early knowledge of basic seamanship—including his remarkable navigation skills—while based in Salem. He also honed his renowned financial and trade calculation skills while growing up in Salem (Latham, 1955). His family had a vital interest in navigation—handwritten navigation manuscripts prepared around 1750 by two of his uncles, Ebenezer and John Bowditch, are held in the Phillips Library (Ellerton & Clements, 2009a). It has been said that during the period 1750–1810 the livelihood of virtually every New England citizen depended to some extent on shipping, and Salem's prominence was the result of creative efforts it made to take advantage of its position by focusing on shipping and trade (Flibbert et al., 1999).

There can be no doubt that the leading merchants of Salem were vitally interested in the quality of mathematics education offered to those likely to be employed in mercantile and navigational work associated with maintaining and extending the work of one of the nation's largest ports and trading centers. A letter signed by eight leading Salem merchants (including William Gray, Nathaniel Bowditch, Jacob Crowninshield, Elias Hasket Derby, Benjamin Hodges and Benjamin Pickman), was included in a list of recommendations for the first edition of Michael Walsh's (1801) *A New System of Mercantile Arithmetic Adapted to the Commerce of the United States*. The recommendation stated:

> We the subscribers, merchants of Salem, convinced of the necessity of rendering the forms of business, the value of coins, and the nature of commerce more familiar to the United States as a commercial people, do approve of the *Mercantile Arithmetic* of Mr Walsh, and recommend it as calculated to subserve in the best manner the instruction of our youth, and the purposes of a well-informed merchant. (p. 5)

Within the set of recommendations for Walsh's book were similar statements made by leading merchants of Newburyport and Boston, both important ports in Massachusetts. Clearly, many merchants greatly valued formal mathematical studies, and many parents were prepared to pay to send their children to private schools rather than to public schools in order that the career prospects of their offspring would be enhanced. A far greater proportion of children in Salem attended private schools than in most other towns (Public Schools in Salem, 1839).

The arithmetic cyphering books in the Phillips Library collection. The 130 cyphering books in the Phillips Library which focus on arithmetic are of special interest for two main reasons. First, there are so many of them; and second, many of them were prepared in or around Salem. This provided us with the opportunity to check our hypothesis that contents of cyphering books would be consistent with local needs. In most of the arithmetic manuscripts there is an obvious emphasis on the arithmetic skills that were needed in the "compting" rooms of any mercantile center and seaport.

The *abbaco* tradition was alive and well in Salem in the 18th and early 19th centuries. Most of the arithmetic problems in cyphering books dealt with practical topics like loss and gain, discount, the various rules of three, currency exchange, compound interest, tare and tret, alligation, fellowship, equation of payments, insurance, barter, and gauging. Those preparing the books were making their own reference books for the business of the wharves. Yet, despite the practical emphases, and the fact that most cyphering books were prepared by boys, the quality of penmanship and calligraphy was usually—but not always—moderately high, for that was part of the cyphering tradition. One boy even wrote: "Excuse my writing, for I had no ruler, nor plummet" (quoted in Ellerton & Clements, 2009a). Some of the cyphering books were produced by sons of prominent Salem families—suggesting that in Salem the tradition whereby children of wealthy families studied mainly the classics had been undermined.

The second point about the Salem arithmetic manuscripts is that there is evidence that in Salem in the late 18th century there was a lot of cyphering done though private tuition (Latham, 1955). John Southwick, a schoolmaster in Salem, prepared "necessary tables for arithmetic" that were printed by "T. C, Cushing, Salem" (Ellerton & Clements, 2009a). Many of the unlined books that became cyphering books were commercially prepared in Salem, and bound within soft marbled covers. By the 1830s, and perhaps before that, cyphering had become part of the curricula at the North, East, West and Central common schools in

Salem, and at least one manuscript was prepared at the prestigious Andover Academy, which was located close to Salem. Girls, as well as boys, prepared cyphering books—for example, Eliza Hook prepared a cyphering book at Salem West School.

Geometry/trigonometry/navigation cyphering books. We studied 65 cyphering books in the Phillips Library that focused on geometry, trigonometry or navigation, and each provided important insights into geometrical and navigation education in the early years of the United States. Throughout the 17th and 18th centuries relatively little geometrical education occurred within colonial America except in brief courses on Euclid and navigation offered by some of the colleges (Ackerberg-Hastings, 2002; Meyer, 1965), and in courses offered by private tutors who focused on preparing boys to become navigators (Simons, 1936a).

It was not until the 1840s that colleges began to place geometry on the list of entrance requirements (Sinclair, 2008). Before that, any young person wishing to study subjects like geometry, plane and spherical trigonometry, logarithms, and the various types of sailing, and mapping, was likely to have had some motivation for doing so. Throughout the 18th century, and into the 19th century, many New England boys aspired to become navigators. These boys knew that if they wanted to be selected as apprentice navigators, they would have to satisfy shipmasters that they had successfully studied the kind of geometry and trigonometry that navigators were expected to know and use. A well-prepared and attractive navigation cyphering manuscript was certainly a step in the right direction.

Although in the 18th century there was no navigation training center like Christ's Hospital, London, in the New World, the influence of Christ's Hospital could be seen wherever navigation classes were held in North America. In the earliest navigation manuscripts held in the Phillips Library collection it is obvious that navigation education was regarded as a serious matter that called for a thoughtful combination of theory and practice. During our first visit to the Phillips Library we identified what we believe is the nation's oldest extant "navigation education" manuscript. It was prepared by an anonymous author between 1692 and 1694, and has 29 pages (dimensions $12''$ by $7.5''$) (Ellerton & Clements, 2009a; Gaydos & Kampas, 2010). Handwritten entries focus on navigation, astronomy, and surveying, with references being made, for example, to the Gunter scale (which had been developed in England in the 17th century).

Of some historical significance is the fact that this 1692–1694 manuscript is mainly in the form of a log of an intended voyage from Salem to Newfoundland. The idea of training potential sailors and navigators to develop a log of a journey was always part of the curriculum of the Royal Mathematical School founded at Christ's Hospital School, from its beginning in 1673 (Allen, 1970; Christ's Hospital, 1953). There are many 18th- and 19th-century handwritten manuscripts held by the Phillips Library in which students recorded logs of actual, or intended, journeys. The 1692–1694 manuscript provides evidence that this same idea was to be found in New England, even before the beginning of the 18th century.

While at Phillips Library we also identified two other very early navigation education manuscripts, one dated 1707 and the other 1711. Although the author of the 1707 manuscript, which occupies just 22 pages and includes some line drawings in ink, could not be identified, the emphasis was clearly on solving geographical, sailing, and astronomy problems. Once again there was a log of a journey. The 1711 manuscript was prepared by Jean le Measurier, and occupied 60 pages, including a log of an intended journey dated from July 11, 1711 (that occurred "by permission of God from Lizard to Boston in New England, in good ship Success of Guernsey under the command of Captain George Guillaume"). It is noted that "on Thursday 29th we had a fresh gale of wind." In fact, the expression "fresh gale of wind" can be found in the logs of hypothetical or real journeys in numerous cyphering books held by the Phillips Library. Altogether, there are seven pages of handwritten notes in the 1711 log relating to this journey, and it is noted that the ship "arrived at Boston by the blessing of God." There are logs for several other journeys, some apparently real, included in this manuscript. The standard of penmanship and calligraphy is medium to high, there being some beautiful line drawings.

Technically, the 1692–1694, and 1707 manuscripts do not fit within our definition of a "cyphering book," for standard IRCEE and PCA genres (as they are defined later in this chapter) were not evident. Nor did their authors attempt high quality penmanship or calligraphic headings. Thus, according to our definition, the earliest cyphering book at Phillips Library is that prepared by Jean Le Measurier in 1711 (Ellerton & Clements, 2009a; Gaydos & Kampas, 2010).

The second-oldest cyphering book of the "navigation" variety held by the Phillips Library was prepared by David Moor between 1727 and 1728. It has 360-pages, and deals with topics like chronology (measurement of time), square and cube roots, straight edge/ruler constructions, right angle and oblique trigonometry, compass variation, various forms of sailing (plane, oblique traverse, middle latitude, Mercator, great-circle). It included logs of voyages, and dealt with astronomy, and gauging. An elaborate, multi-colored compass was drawn, and creative, "dynamic" diagrams (with moving parts) constructed. This manuscript has a strong Christ's Hospital feel about it, and the log of a voyage—almost certainly not an actual voyage—from the Lizard to Barbados ("begun Feb 1st, 1727") was presented.

There are many other 18th-century "navigation" cyphering books in the Phillips Library collection, with dates spread throughout the century. The most impressive of all of them is a magnificent 360-page manuscript prepared by Henry Tiffin at various times between 1758 (when Henry was 10-years-old) and 1774. This manuscript features numerous water-color paintings. The mathematics covered in the book involved straight-edge/compass constructions, trigonometry, logarithms, sketching profiles of coastlines, and various kinds of sailing (plane, oblique, Mercator, and middle-latitude). Many of the paintings seem to have been done on board vessels, when Henry was serving as an apprentice navigator (Ellerton & Clements, 2009a).

Almost all the navigation cyphering books held in the Phillips Library have a similar content structure. The first topic would be concerned with Euclidean descriptions or definitions of plane geometrical figures (for example, "A point is that which has no parts, being of itself indivisible"); then would follow theorems, proofs, and the time-honored compass/straight edge constructions. Despite Nathalie Sinclair's (2008) statement to the contrary, our analysis of the manuscripts made it clear that these would-be navigators did, in fact, use straight edges and compasses and actually carried out the constructions. Arcs from compass constructions remain clearly visible in most of the navigation manuscripts. Learning to be a navigator involved much more than mere memorization, and when examining the navigation cyphering books we always got the impression that students had been actively engaged in the learning process.

Once the elementary geometry sections, with their straight-edge and compass constructions, were completed, then an introduction to logarithms and plane trigonometry would follow (Waters, 1958). By this stage, the mathematics was getting more serious, and would have been too difficult for those who had not gone beyond an elementary course in arithmetic (up to the single rule of three). Only the dedicated or the ambitious moved on to this next stage, in which they had to make sense of the standard, but rigorous, studies of the different types of sailing (plane, oblique, traverse, middle latitude, Mercator, great circle), and methods for taking lunar observations, and plotting daily positions of a ship, and oblique and spherical trigonometry. Finally, navigation students needed to learn how to keep a log in which wind and current directions and speeds were recorded, and this was done through making logs of real or imagined voyages. As indicated, previously, the 1692–1694 navigation manuscript held by the Phillips Library is essentially a log of an imagined journey from Salem to Newfoundland.

As noted previously, Nathaniel Bowditch (1773–1838), arguably one of the greatest navigators the world has ever seen, was a Salem man. Between 1795 and 1803 he made five voyages on merchant ships to the East Indies and China, serving on the last voyage as master and part owner (Latham, 1955). His classic *American Practical Navigator,* first published in 1802, would become known, simply, as "the Bowditch." It is still recommended for reading in navigation education courses conducted by the United States Navy. On a trip from Salem to the East Indies, Bowditch apparently taught every member of his crew of 12, including the cook, "to take and calculate lunar observations and plot the daily position of the ship" (United States Hydrographic Office, 1943, p. 3).

What were the origins of this approach to navigation education? It would be wrong to call it the Salem, or the Bowditch, approach, for a similar approach was recommended by Archibald Patoun in his classic *A Compleat Treatise of Practical Navigation [Demonstrated from its First Principles; Together with all the Necessary Tables. To which are Added, the Useful Theorems of Mensuration, Surveying, and Gauging,* the first edition of which was published in London in 1730. The 29-page entry on "Navigation" in the first edition of the *Encyclopaedia Britannica* (1771), published in Edinburgh, Scotland, was based on a later edition of Patoun's much-published work. The last two pages of the *Encyclopaedia Britannica* article show sample log book entries for a journal "from the Lizard towards Jamaica in the ship Neptune" (p. 393). Interestingly, many of the logs of journeys recorded in navigation manuscripts in the Phillips Library began from "The Lizard," and the recommended structure of the log, and the forms of language used, were very similar to what is found in the *Encyclopaedia Britannica* article in 1771. "The Lizard" is the most southerly point of the mainland of Great Britain.

We believe that the basic structure for the approach to navigation education found in the Salem navigation cyphering books came from England, and in particular, from the Christ's Hospital navigation program. As noted previously, in the early years of its existence, the Christ's Hospital Royal Mathematical School scholars based their cyphering books on a textbook prepared, especially for the school, by Sir Jonas Moore and Peter Perkins (1681). In 1709, a complex work on algebra, originally written in Latin, by the Swiss mathematician Johannes Alexander, was printed at the expense of the governors of Christ's Hospital. This text had been translated into English by Samuel Cobb, a master at Christ's Hospital, and 750 copies of the Latin edition together with 1000 copies of the English translation were printed. A 128-page appendix, written by Humfry Ditton (1709), a master at Christ's Hospital, was added to Alexander's (1709) original text.

In the early 1720s, James Hodgson, Master-in-Charge of the Royal Mathematical School at Christ's Hospital between 1708 and 1754, committed the school's navigation curriculum and methods to two large volumes (Allen, 1970). After informing his readers on the title page of Volume 1 that he was "Master of the Mathematical School in Christ's Hospital, and Fellow of the Royal Society," Hodgson (1723) wrote in his Preface that the volumes set out the methods he used in teaching navigation, and the rules by which he himself operated. He continued:

> I have endeavoured thro'out the whole course of this work to deliver everything with as much plainness as possible. ... I can't recollect that there is one thing left undemonstrated, that is capable of it. ... the calculations are all done to the greatest degree of exactness, and have all gone through my own hands. (pp. vii–viii)

The publication of Hodgson's volumes meant that a text setting out the Christ's Hospital curriculum and methods for navigation education became available to all. The influence of Christ's Hospital would be felt in the American colonies (Allan, 1984; Brooks, 2010; Coldham, 1990), and when examining navigation manuscripts at the Phillips Library we found evidence that the curriculum and methods established by Christ's Hospital had been followed in the New World. Let us look briefly at that evidence.

The sequencing of the topics in the Salem navigation cyphering books was almost identical with that in Hodgson (1723), and in handwritten navigation cyphering books prepared by Christ's hospital students of the eighteenth and early 19th century (e.g., Anderson, 1840; Andrew, 1836; Janeway, 1767; Mott, 1837; Page, 1825; Raffles, 1755; Stokoe, 1790; West, 1785). Numerous examples presented by Hodgson (1723, Vol. 1, pages 237, 244, 245, 246, 247, 250, 251, etc.) involved a ship sailing from "the Lizard." Between pages 417 and 433 of Volume 1, Hodgson (1723) showed a log of a journey in 1721 that started at The Lizard. Every day, entries mentioned the present bearing of the ship, and its distance from The Lizard. In many of the logs in the Phillips Library navigation cyphering book collection, the wording of corresponding sections was the same, or almost the same—but the dates listed were different. Quite a few of the journeys were described as having started at "The Lizard." It is unlikely that any of the logs were prepared during actual trips.

A log for a journey between Salem and Charlestown (South Carolina), kept by Ebenezer Bowditch, in his 1749–1751 navigation cyphering book, contained idiosyncratic descriptions, and was probably based on an actual journey. By contrast, log books shown in manuscripts prepared by Jean Le Measurier in 1711, and David Moor in 1728, and indeed in numerous other navigation manuscripts held at the Phillips Library, referred to journeys that started at The Lizard. The books prepared in London by Thomas Stokoe (1790), James Janeway (1767) and Charles Page (1825), three Christ's Hospital students, had logs in which the destination was Madeira, and in many of the Phillips Library navigation cyphering books prepared after Hodgson's period, Madeira was also the destination. For navigation cyphering books prepared at Salem, often the starting point was Boston (or some other North American port), and Madeira was the stated destination. The evidence supports the conjecture that in Salem (and in other places in North America), the curriculum and the methods for training navigators were consciously modelled on the Christ's Hospital (or the Royal Naval Academy, at Portsmouth) curricula and methods.

There were major structural differences, however, between the Salem navigation training programs and the Christ's Hospital program. At Christ's Hospital the boys studied full-time for at least five years, and sometimes for up to ten years (Christ's Hospital, 1785; Trollope, 1834), but the navigation classes in Salem were conducted, typically in one-on-one tutorial sessions, or in small groups—usually in small private schools during evenings. Edmund Knight, of Salem, mentioned in a cyphering book prepared during 1763–1764, that he had had 50 lessons from "Master Moody," 45 on arithmetic and 5 on geometry, and that the total tuition cost was 2 pounds, 14 shillings and 6 pence. The Salem navigation teachers knew that their main task was to offer tuition which complemented what their students were learning, or had already learned, as a result of their daily activities in or around Salem.

The Salem students probably did not have access to James Hodgson's (1723) volumes or, for that matter, to other printed textbooks—more likely, any text to which they referred would have been in the form of handwritten manuscripts passed on, or sold, to them by others who had completed the training. Thus, for example, the navigation cyphering book that Jacob Morgan prepared in 1773 includes a note that it was sold to Wesley Burnham in 1783. Burnham would subsequently become a well-known captain of ships that traded between Salem and the East Indies.

Throughout the period 1780–1830 many Harvard College students took courses in navigation. They too produced navigation cyphering books that reflected, in their content and sequencing, the Christ's Hospital model. In May 2009 and July 2010 we examined 14 navigation cyphering books held by the Houghton Library at Harvard University and almost all of these featured high-level penmanship, calligraphy, and line drawings (some had water-color illustrations). We concluded that 6 of the 14 manuscripts were originally prepared in Europe (in Great Britain or in France), and the remainder were prepared by American students.

Because Harvard College largely followed the Christ's Hospital model, the Harvard navigation cyphering books reveal a curriculum that attempted to balance theory with practice. This can be seen by examining navigation cyphering books, held in the Houghton Library, prepared by William Hancock in 1722, William Winthrop between 1769 and 1774, John Williams in 1779, Joshua Green in

1782, Joshua Bates in 1798, and C. Lowell in 1798 and 1799. Hancock's (1722) manuscript featured a log of a trip from Ireland to Virginia.

William Roberts's (1742) 341-page manuscript included a log of a journey that although dated 1738, probably never actually took place. This manuscript was not prepared at Harvard College, for an inscription indicated that Roberts attended the "Ron Academy" (which almost certainly referred to the "Royal Naval Academy" founded in Great Britain in 1733). A 37-page manuscript by Moses Mowry (c. 1780), which included a log of a journey from England towards Madeira, is held in the Houghton Library collection. This manuscript was prepared at a school in England, and its content suggested that the school was following a navigation curriculum based on that at Christ's Hospital.

In the 1930s, Lao Genevra Simons (1931, 1936a) studied geometry curricula and instruction methods at Harvard, Yale, Columbia, and Philadelphia colleges/universities during the colonial and early Federal periods. She concluded that although many students were introduced to Euclid's *Elements,* more attention was given to what she termed, somewhat disparagingly, "practical geometry" than to "Euclid" (p. 604). When discussing a 75-page notebook on practical geometry prepared by Samuel Miller in 1788, at the University of Philadelphia, Simons (1936a) made clear her belief that, in comparison with Euclid, the educational merit of navigation and surveying education was questionable. She wrote:

> It includes no definitions, no references to Euclid ... It consists of a set of constructions given without description and proof, such a set as was once presented in a course in so-called "mechanical drawing." ... Every one of these "Practical Geometry" notebooks concludes with sections on some or all of the subjects trigonometry, dialling, navigation. (p. 604)

Smith and Ginsburg (1934) stated that early geometry education in the United States was "merely an aid to the study of astronomy" (p. 13). Sinclair (2008) repeated this assertion. According to Sinclair (2008), the geometry curriculum in 18th- and early-19th-century North American colleges "*was* literally the *textbook* [original emphasis], with university students memorizing one proposition after another in the given sequence" (p. 15). However, Sinclair's statement is not consistent with Colyer Meriwether's (1907) view that content in textbooks and instruction in school and college classrooms were often only superficially similar to each other, because textbook authors were too far removed from the worlds of students who might be expected to use their texts.

Simons's (1936a) and Sinclair's (2008) studies of geometry in colonial colleges or universities indicated that relatively few colonial students seriously studied theoretical geometry. But the manuscript evidence we have examined made clear that all students who studied astronomy or navigation at Harvard, or in private classes in Salem, did pay much attention to practical geometry associated with navigation and surveying. And, almost all of them did address Euclidean definitions and axioms, as well as to Euclid's elementary straight-edge and compass constructions. Simons (1936a), Sinclair (2008), and Smith and Ginsburg (1934) seemed to believe there was little education or practical merit in such studies, but we would disagree with their value judgments on that issue.

Practical Emphases in Arithmetic and Navigation Cyphering Books 69

In a later section we provide evidence that aside from the relatively small amount of Euclidean geometry—comprising definitions and straight edge/compass constructions—the main form of geometry studied by colonial boys was the practical geometry expected of boys preparing to be navigators or surveyors. In most cases, this practical geometry was taught by private tutors or by instructors in academies that increasingly were to be found in major towns during the 18th century (Eby & Arrowood, 1934).

Sinclair's (2008) statement notwithstanding, most students preparing to be navigators, or surveyors, or even those preparing for college navigation courses (such as those offered at Harvard College) usually did not own a commercially-published textbook on geometry, or navigation, or surveying. As previously argued, the "textbooks" used in these courses were likely to have been earlier handwritten navigation (or surveying) cyphering books—perhaps owned by the school or college, or by instructors, or perhaps passed on within families from generation to generation (Van Sickle, 2011). Commercially-published, printed textbooks on geometry were expensive and, in any case, not entirely necessary under the cyphering tradition.

Direct, whole-class teaching to classes of students did not occur. Private tutors and academies offered practical classes for the benefit of indentured workers and any other interested persons. In New England, these classes usually took place throughout the winter months or in evenings. An individualistic, applied form of pedagogy was adopted, with male students learning reading and writing, cyphering, practical geometry (constructions with straight edge and compass, and the geometry of the sphere). Some young men studied one or more "applied" subjects—such as bookkeeping, logarithms and elementary trigonometry, the various types of sailing, and surveying. Female students rarely studied these subjects—they tended to concentrate on English grammar, sewing, cyphering, bookkeeping, a modern language (especially French), housekeeping, and accomplishments such as elocution and dancing (Eby & Arrowood, 1934).

Often, students would take only those courses that were directly related to their everyday workplace affairs. Thus, for example, those who had a daily responsibility for looking after a storekeeper's accounts would be likely to take courses in writing, cyphering, and bookkeeping (see, e.g., Nichols, 1921). Subjects like algebra, and formal Euclidean geometry, were usually only studied in higher-level colleges like Harvard and Yale. The popular attitude to what comprised mathematics was summed up in the following excerpt from Charles Peirce's (1806) book, *The Arts and Science Abridged:*

> Q. What is meant by the mathematics?
> A. A science that contemplates whatever is capable of being numbered or measured. It ranks the first of all sciences, because it consists only in demonstrations.
> Q. How are the mathematics divided?
> A. Into arithmetic, geometry, architecture, astronomy and mechanics. ... (p. 43)

Notice that Peirce did not mention algebra or trigonometry. Peirce placed a particularly high value on the educational and practical merit of arithmetic, which he regarded as something every person should learn because it was "the soul of commerce, and the mother of all sciences" (p. 43). He believed that although all children should begin to cypher at around the age of 9 or 10, there was "no purpose for them to begin younger" (p. 45).

Genre considerations. We have argued that the genre of navigation cyphering books prepared in colonial North America, and in the early Federal period up to the middle of the 19th century, was largely determined by a desire to imitate the genre that the Royal Mathematical School at Christ's Hospital, in London, required of students who were preparing handwritten volumes that they would take with them when they graduated to apprenticeships as full-time navigators. This Christ's Hospital genre had been defined by a committee associated with the Royal Mathematical School in the late 17th century, and such was the success of the School that it acquired the reputation as the leading navigation training school in the world.

The genres within the Christ's Hospital manuscripts were similar to those genres of standard *abbacus* cyphering books which incorporated what we have called the IRCEE (Introduction-Rule-Case-Example-Exercises) and PCA (Problem-Calculation-Answer) genres, but there was one noticeable difference. Although it is true that the *abbaco* IRCEE genre can be found in every Christ's Hospital handwritten volume, and in every navigation cyphering book prepared in North America that we have examined, a modified form of PCA genre was evident in the Christ's Hospital handwritten volumes. The modification was that although written explanations for rules ("reasons for rules") were clearly shown in most Christ's Hospital manuscripts, that was not the case with most North American manuscripts.

Ethnomathematical and Methodological Considerations

From a mathematics education perspective, the manuscripts produced by students in Salem during the 18th and early 19th centuries testify to a special situation. Salem was a prosperous seaport, and it was normal for Salem boys to work on the wharves, or to work in trading stores, or to be apprenticed to coopers. The world in which commodities such as flour, pepper, gunpowder, tobacco, molasses, rum, wine and milk, were traded was very familiar to them. In Salem, for example, casks were custom-built to serve as effective shipping containers, and it was natural that coopers would be interested in the arithmetic of gauging, by which the capacities of casks could be measured. Nathaniel Bowditch's father was a cooper, and Salem boys became apprenticed to coopers; others were trained, during their apprenticeships, to be reckoners and book-keepers (Latham, 1955). Masters regarded it as important that their apprentices became efficient at making dock-side calculations involving concepts like tare and tret. Those who worked in offices in which partnerships for forthcoming trade ventures were negotiated needed to learn the principles of the arithmetical topic known as "fellowship."

As a natural part of the process of growing up in Salem, boys learned about seaport-related arithmetical activities such as alligation, gauging, fellowship, tare and tret, barter, profit and loss, currency exchange, mensuration of 2- and

3-dimensional figures, compound operations, reduction, and ratio and proportion (the "rules of three"). Boys were on the wharfs when sloops returned from Sumatra laden with precious, exotic spices, and they witnessed, and participated in, the action when precious cargoes were measured, taxed, packaged and sold. Most of these activities needed to be recorded in concise, arithmetically accurate forms (Latham, 1955).

An unusually high proportion of Salem boys aspired to go beyond the world of mere calculation: they wanted to become navigators, for that risky profession seemed to promise a life of action on the high seas and adventures at the edges of the known world. We identified more navigation cyphering books in the Phillips Library at Salem than in all of the other collections that we have seen, combined.

Thus, the concepts and skills that were summarized in cyphering books prepared in Salem and in other New England ports such as Newburyport would have had meaning for those who prepared them. What students wrote in their cyphering books was associated with thought processes that were embedded in reality. The sensory, mental and physical efforts they expended on such topics as alligation would have developed strong links between theory and practice (Bregman, 2005; Jonassen & Rohrer-Murphy, 1999).

Modern mathematics education researchers use the term "ethnomathematics" when they examine mathematical aspects of everyday behaviors of community and family members. Those who research ethnomathematics study the shared mathematical language, knowledge, skills and preferences of those who grow up within well-defined communities (Babu, 2007; D'Ambrosio, 1986; Emigh, 2002; Lave & Wenger, 1998). They recognize that what is "normal" in one community, geographical region, trade, or profession, is often given less attention in another.

Unlike the situation in Salem, most boys growing up in New York in the 18th century did not yearn to work on the wharves as reckoners, or to become navigators guiding ships to the East Indies. They were more inclined toward becoming secretaries, or clerks, and New York families believed that arithmetic "up to the rule of three direct" was all that would be needed so far as mathematics was concerned. According to Karpinksi (1926), this resulted in early arithmetics printed in New York being used so much that they were discarded, and therefore became scarce. What was true of New York was also true, to a certain extent, of Boston and Philadelphia. But even students who had no access to textbooks were expected to prepare cyphering books. Around 1810, for example, students attending public schools in Boston were expected to cypher up to and including the double rule of three, direct and inverse (White, Burditt & Co., 1809).

Benjamin Franklin did not learn arithmetic well when, aged between 8 and 10 years, he cyphered at the Boston Latin School. Subsequently, however, he learned arithmetic associated with the trade of printing, when he was apprenticed to older brother, James, a London-trained printer (Best, 1967; Randall, 1975; Thorpe, 1893).

Figure 3.7 shows Thomas Rowlandson's depiction of *A Merchant's Office in London* at the end of the 1780s. The ability to enter commercial records accurately and neatly was regarded as an important part of office work, and thus it was only natural that writing-as-also-arithmeticke emphases were important in schools preparing

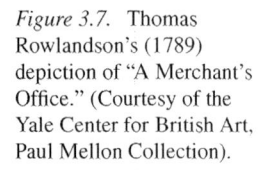

Figure 3.7. Thomas Rowlandson's (1789) depiction of "A Merchant's Office." (Courtesy of the Yale Center for British Art, Paul Mellon Collection).

children for office work. If a boy wanted to go into business then the first step was likely to be an apprenticeship in a merchant's office, where good handwriting and reckoning skills were needed.

From an ethnomathematical perspective, we needed to examine as many cyphering books as possible that had been created in different places and different contexts. Undoubtedly, ethnomathematical expectations were behind the high quality penmanship and calligraphy evident in many of the cyphering books that we have examined (Thornton, 1996). In addition to the 212 cyphering books in our own collection, we studied those in the Phillips Library, as well as those in smaller, but still significant, collections of cyphering books in the Houghton Library at Harvard University, the Clements Library at the University of Michigan, the David Eugene Smith Library in the Butler Library at Columbia University, the Thomas P. Hill collection at the University of California, the Wilson Library at the University of North Carolina, the New York Public Library, the American University Library (Washington, DC), and the Swem Library at the College of William and Mary (Williamsburg, Virginia). Online descriptions of the cyphering books in the special collection held by the Huguenot Historical Society, in New Paltz, New York (see Appendix C) were also examined.

Cohen (1982) stated that there were about 60 cyphering books in a collection in the Harvard University libraries. We found about 80 there, but many of them (especially those in the Houghton Library) originated from Europe, and not from the United States of America. There is a collection of at least 65 cyphering books held in the Wilson Library at the University of North Carolina at Chapel Hill, and almost all of these originated from southern states, as did the 28 cyphering books that we examined in the Swem Library within the College of William and Mary. When we examined the collection of between 35 and 40 cyphering books held in the Clements Library, at the University of Michigan, we found that although these originated from various parts of the United States, they were often associated with persons who had been influential in the Revolutionary War of the 1770s and 1780s. There are about 25 cyphering books—originally prepared by persons living in Huguenot families in the eighteenth and 19th centuries—held by the Huguenot Historical Society, 18 Broadhead Avenue, New Paltz, NY. There are about 20 miscellaneous cyphering books in the Thomas P. Hill collection of early American mathematics book, held by the University of California, and most of these originated in the United States.

Of special interest to us was the fact that we found cyphering books prepared by students attending the Royal Mathematical School, Christ's Hospital (London) in the 18th century, held in the Phillips Library, in the Clements Library (at the University of Michigan), in the Butler Library (at Columbia University), and in the New York Public Library. We have another two in our own collection.

From a research design point of view, it was critical that we addressed the delicate issue of how a representative sample of cyphering books might be obtained. In 2005 we set out to do this by using our own resources to purchase as many cyphering books through on-line auctions as possible. We have succeeded in establishing a collection of 293 handwritten books (of which, 212 were prepared in the United States and satisfy our criteria for what constitutes a cyphering book). The story of how we did that, and how that collection enabled us to provide what we believe are strong answers to our original research questions, will be given in remaining sections of this work.

Comments on the Effectiveness of the Cyphering Approach

Throughout the 17th and 18th centuries, and during the first half of the 19th century, most school authorities in European nations, and in those American colonies which would become part of the United States of America, believed that the best way for children to learn mathematics was for them to write definitions, rules, cases, type problems and solutions, and accurate solutions to additional exercises, into personal cyphering books. For most of the 18th century the model problems and exercises were taken from cyphering books owned by teachers, but toward the end of the 18th century, and into the 19th century, they increasingly came from printed textbooks (Monroe, 1917).

In many—perhaps most—early North American schools, arithmetic was part of the curriculum. It was taught through a cyphering approach, and was typically a component of the writing program (Monaghan, 2007). In Boston in 1789, for example, it was ordered that "three reading schools and three writing schools be established" for the instruction of children between the ages of 7 and 14. Whereas the reading schools taught spelling, accentuation, reading of prose and verse, English grammar and composition, the only two subjects taught in the writing schools were writing and arithmetic (Cubberley, 1920).

With the benefit of hindsight, historians have almost universally condemned the cyphering approach to mathematics education (see, e.g., Cohen, 1982, 1993, 2003;

De Morgan, 1853; Meriwether, 1907; Williamson, 1928; Yeldham, 1936), claiming that it over-emphasized drill and practice, placed too much stress on memorization, and wasted too much time on mindless copying. These and other historians (e.g., Small, 1914; Van Sickle, 2011) have emphasized the importance of commercially-printed textbooks in the evolution from what they regarded as the barren cyphering approach to more satisfactory interactive approaches (see, for example, University of Michigan, 1967). However, these historians have not fully addressed an associated question: If the cyphering approach was so unsatisfactory, why did it endure for over 600 years throughout a period which was marked by such rapid mercantile, industrial and scientific advances?

Critics of the cyphering approach have alleged that even though the content in many of the cyphering books appeared to be "practical," much of it was irrelevant to the students' immediate and future lives (e.g., Cohen, 1982, 2003). These critics have further maintained that the content was too difficult for most learners, that the teachers themselves rarely had any understanding of what their students wrote, and that, in the process of cyphering, students rarely became actively engaged, mentally (see, e.g., Cajori, 1890, 1907; Cohen, 1982, 2003; De Morgan, 1853; Halwas, 1997; Middlekauff, 1963; Monroe, 1917; Yeldham, 1936). According to most commentators, students were merely expected to memorize the material they copied and to prove that they had done so by reciting what they had memorized in follow-up recitation sessions. This was the case, it was sometimes claimed, in colleges as well as in schools (see, e.g., Wayland, 1842). Richard Hofstadter and Walter Metzger (1955) claimed that students found recitation "tiresome" and teachers found it "stultifying" (p. 229).

Simons (1936a, 1936b) provided evidence of the acceptance, even by college students, of the view that, above all else, a good teacher of mathematics was able to get students to memorize important excerpts from mathematics texts. Simons (1936a) referred to an incident in connection with the appointment to Columbia College in 1786 of a certain John Kemp, aged 23, as Professor of Mathematics. In 1785 Kemp had been provisionally appointed to teach mathematics for one year at Columbia College, and at the end of the year a public examination of his class was held, in which each student was required to draw a number out of a box and demonstrate without further assistance the problem or theorem in Euclid to which it referred. The students did this very convincingly, and because of that the College authorities appointed Kemp to the position he sought. Stanley Guralnick (1975) indicated that this attitude was generally accepted in the American colleges of the 18th century. Professors worked from their own cyphering books, which students copied to produce their own cyphering books (Van Sickle, 2011).

According to Cohen (1982), an over-emphasis on copying and memorization resulted in easy topics taking on "nightmarish qualities" (p. 121). Teachers carried heavy workloads, irrespective of the levels at which they taught, and cyphering provided no relief. Furthermore, there were relatively few books in college or school libraries for students to read. As late as the 1830s, for example, Yale College had a total of only 8500 volumes in its Library (Davenport, 1832, p. 25). The cyphering/recitation approach helped teachers to maintain discipline. Often, students were not allowed to make entries in cyphering books unless they had first memorized what they would write (Hofstadter & Metzger, 1955).

Cohen (1982) is among many to have questioned the relevance of the content and forms of mathematics that students were expected to learn. She wrote:

> Arithmetic was a commercial subject through and through and was therefore burdened with the denominations of commerce. Addition was not merely simple addition with abstract numbers, it was the art of summing up compound numbers in many denominations—pounds, shillings, pence; gallons, quarts, and pints (differing in volume depending on the substance being measured), acres and rods, pounds and ounces (both troy and avoirdupois), firkins and barrels, and so on. Eighteenth-century cyphering books show that students had to memorize all these tables of equivalences before embarking on the basic rules and operations. A large chunk of time was spent on a subject called "reduction"—learning to reduce a compound number to its smallest unit in order to facilitate calculations. Students would practice on questions like "How many seconds since the creation of the world?" and "How many inches in 3 furlongs and 58 yards?" This would prepare them for more advanced problems, such as "What will ten pairs of shoes cost at 25s 6d a pair?" (p. 121)

According to Cohen (2003), the cyphering approach "involved transcribing formulaic rules out of printed textbooks, word for word into the pages of individual books, together with the exact examples offered by the text to illustrate each rule" (p. 44).

Although we (Ellerton and Clements) acknowledge the validity of the point Cohen made about difficulties caused by the large number of units for measurement, we nevertheless think that in the above passage she over-stated her case. Over the past six years we have studied about 600 North American cyphering books, each prepared during the period 1701–1861, and have generally found the content to be well sequenced. We have observed only a few instances when "applied operations" were dealt with before the four operations on abstract numbers. Many students wrote down tables of equivalences, but this was not usually expected of them before they had dealt with the basic rules and operations.

Despite some historians' assertions to the contrary (e.g., Cohen, 2003), we do not believe that most cyphering books were generated by students who copied long sections from printed textbooks. Reminiscences of those who created cyphering books suggest that most of the material that they copied came from pre-existing handwritten cyphering books or from their own draft problem solutions, written on scraps of paper (or, perhaps, slates). Often, draft solutions were checked for their correctness by a teacher or an "usher" before students neatly copied them into their cyphering books (Lancaster, 1805).

Most pertinent of all, our analyses of several extensive cyphering book collections held in libraries across the United States suggested to us that, sometimes, what students copied into their cyphering books was regarded as highly relevant to their present situations, and in line with their aspirations for the future. Whether that was *always* the case was beside the point—those who prepared cyphering books, and future prospective employers, *believed* that the process of preparing the books was an important, indeed essential, step towards professional competence.

Thomas Burrowes (1862), Superintendent of Pennsylvania's Common Schools in 1861, argued that, during the first half of the 19th century, the proximity of common schools in New England states to seaports meant that it was natural, in those states, for there to be an emphasis on navigation and maritime-related arithmetic in the curricula of common schools. In Pennsylvania, however, most of the children in the common schools were from scattered farms, and Burrowes believed that this should have been reflected in school curricula more than it was. In the Southern States it was not expected that the children of slaves would attend school, and school curricula tended to be directed at the needs of the well-to-do (Burrowes, 1862; Eby & Arrowood, 1934; Meyer, 1965).

Burrowes (1862) also contended that early colonists in Pennsylvania originated from many parts of Europe, and "were amongst the most intelligent of the age" (p. 251). This made it necessary that those designing curricula kept their eyes on the entrance requirements of colleges—in other words, the classics and arithmetic were especially important (Hornberger, 1968).

The Evaluative Function of Cyphering Books

The evaluative function of cyphering books appears to have escaped the notice of historians. The idea that both the efficiency of teachers and the quality of student learning were reflected in what could be found in the students' cyphering books was inherited from Europe. Thus, for example, once every year boys in the Royal Mathematical School at Christ's Hospital, in London, were expected, personally, to present their cyphering books to the reigning monarch. William Trollope (1834) wrote that on these occasions the boys produced their maps and charts, and other specimens of their proficiency in nautical science, and, kneeling on one knee, they unfolded these before the king, as he passed. Each was addressed in turn, "and every breast beats high in acknowledgement of the honour conferred by the notice of the Sovereign" (p. 93).

This idea of the quality of learning and teaching being assessed through the visitation of some higher authority was common in grammar schools and colleges in England in the 17th century, and was translated into the North American colonies (Connor, 1902; Eggleston, 1871; Knox, 1788; Watson & Kandel, 1911). At the end of a term of work (which usually comprised about 12 weeks), in local schools in the American colonies, a small committee, made up of professionals such as clergymen, physicians, and lawyers, would visit the school and, in front of parents and friends, would proceed to address questions to individual students.

One after the other, students would stand, and visitation committee members would ask them questions. Each student was expected to answer without help but, in practice, the teacher, who would nervously stand next to the committee, would often rephrase questions in order that they might be better understood. After students had answered the questions, the teacher was usually expected to make a brief statement about what had been achieved at the school during the semester. If a parent dared, he or she might ask the teacher a question. Then might follow choral work by the children and, finally, afternoon tea or supper would bring the event to its conclusion—although, during that time of apparent informality, cyphering books, sewing samplers (Ring, 1993) and other student-generated artifacts would be inspected, both by parents and by members of the visitation committee (Connor, 1902; Eggleston, 1871; Johnson, 1907).

Although everything was usually conducted in a polite way, local politics sometimes intruded (see, e.g., State of Massachusetts, 1848). Local communities were small and, often, students and their parents were well known to the visitation committee. In 1826, John Tinkham (Appendix A, m/s #123) completed entries in his cyphering book with the following comment:

> Here you see the end of vulgar fractions, and the end of the book. I think if the dam'd committee looks in this book they will find out just what I think about them. The names of those dam'd rascals are thus: Daniel Davis, Captain Gordon and Doctor Mahew, one of the dam'd Presbyterian mohogs. I s'pose you have seen the man. He rides a white horse with a black spot under his tale, and when he rides his back goes before he waves a bobbtale (sic.) coat, a mouse skin cap, and a pair of trousers one leg shorter than the other. This is the dimensions of the doctor.

Tinkham took a risk in writing such a comment because it was important, for both students and their teacher, that the cyphering books made as favourable impression as possible on the visitors and parents. Appearance was everything, and a teacher's chance of further employment as a teacher hinged, to a certain extent, on what committee members and parents thought of the cyphering books and sewing samplers. This, together with the fact that a teacher of arithmetic was expected to double as a writing teacher, helps to explain why so many students and teachers spent much time making the pages in cyphering books as attractive as possible.

Throughout the cyphering-book era it is likely that most parents and friends who attended special end-of-term occasions, and even some members of the visitation committees, could not follow much of the arithmetic in the cyphering books. The cyphering books therefore needed to *look* good, and it is not surprising that some teachers not only ruled the pages of their students' cyphering books, but also wrote the headings, often using elaborate calligraphy (Burtt & Davis, 1995). The examiners and parents expected penmanship, calligraphic headings and diagrams to be of a high order. Discipline in the schools was, more often than not, unduly harsh and in many accounts of early schools (e.g., Dickens, 1850; Nichols, 1921) students were thrashed because material they had entered into their cyphering books was "not good enough."

Cyphering books also provided valuable credentialing evidence. After about 1750, most colleges stipulated that those seeking admission had to demonstrate knowledge of elementary arithmetic. During the last two decades of the 18th centuries the North American colleges began to exert strong downward pressure on the schools and academies so far as mathematics entrance requirements were concerned. Table 3.1, which is based on a table prepared by Edwin Broome (1903, p. 39), shows the college-admission mathematics requirements for entrance to Harvard College in 1798, to Yale College in 1800, to Princeton College in 1794, to Columbia College in 1786, to Brown University in 1793, and to Williams University in 1795.

Table 3.1
Comparative Table of College Mathematic Entrance Requirements, 1786–1800

Institution	Location	Year	Mathematics Entrance Requirement
Columbia College	New York	1786	Arithmetic, to the rule of three
College of Rhode Island (later Brown University)	Providence, RI	1793	Rules of vulgar arithmetic
College of New Jersey (later Princeton University)	New Jersey	1794	Arithmetic (to reduction)
Williams College	Williamstown, MA	1795	Rules of vulgar arithmetic
Harvard College	Cambridge, MA	1798	Nil
Yale College	New Haven, CT	1800	Rules of vulgar fractions

With respect to Table 3.1 it should be recalled that the universities did not set written entrance examinations. A person wishing to enter college would appear before a small committee, probably made up of the college president and several tutors, and would be expected to respond verbally to questions asked by the committee. So far as arithmetic was concerned, the presentation of a very neat, and convincing personal cyphering book to the committee was likely to be an important aid to being admitted (Broome, 1903; Brubacher & Rudy, 1958). This continued until the 1840s, when colleges began to set written entrance examinations for students.

Cyphering books might also be presented to potential employers as evidence that an applicant had good penmanship and arithmetic skills (e.g., Ambulator, 1780).

The key question, which deserves more attention than it has been given by previous historians of mathematics education, is this: Did the cyphering approach help students to learn the mathematics that they would need in order that they would be able to survive with dignity in their present and future lives? In the absence of a large body of data generated by written examinations, and by relationships between the quality of students' cyphering books and key aspects in the students' subsequent lives, it is difficult to answer that question objectively, but we will return to it in a later chapter.

Chapter 4
Formulating the Research Questions

Abstract This chapter provides a personalized account of how and why the authors (Ellerton and Clements) became interested in cyphering books. It also tells how the authors came to believe that the early history of school mathematics in North America between 1607 and about 1861 needed to be rewritten—because previous historians had not recognized the fundamentally powerful influence of the cyphering tradition. The chapter closes with a statement of five research questions.

Our Journey

Before listing a set of questions that guided our research it will be useful to describe how and why we decided to investigate issues associated with the cyphering tradition.

For many years, both of us (Ellerton and Clements) have had a professional interest in the history of mathematics education (see, for example, the emphasis on historical matters in Clements & Ellerton, 1996). We first met in 1985, but even before that both of us had established collections of old mathematics textbooks. We combined our separate collections in 2005, and our mutual interest in old textbooks led us to try to coordinate, understand, and further develop our collection. We also decided to try to purchase additional books, manuscripts, and artifacts so that we could develop a more coherent and complete collection.

We began buying very old mathematics textbooks through online auctions. After purchasing, and studying, Louis Karpinski's (1980) monumental guide to early (up to 1850) American mathematics texts, we decided that we would aim to purchase at least one original (preferably, but not necessarily, first-edition) copy of every North American mathematics textbook listed by Karpinski. At the time of writing, we are close to having achieved that aim. We have a collection of about 1500 different titles listed by Karpinski—of those about 400 are first editions. But, although we have retained an interest in these textbooks, and still work toward achieving our original aim, we soon found our major attention directed at a different but related aim. Our new focus of attention became cyphering books.

We became aware that old handwritten books—variously termed "cyphering books," "ciphering books," "copybooks," "copy books," "exercise books," "note-books," "workbooks," "journals"—were occasionally being sold through on-line and other auctions, and we decided to purchase as many of them as we could in an attempt to become more cognizant of their historical significance. After purchasing a few of them we observed that the penmanship and calligraphy were often of a high standard, and that most were primarily concerned with elementary arithmetic. Some of the manuscripts, however, focused on more advanced arithmetic, algebra, geometry, trigonometry, navigation or surveying.

We observed that in the manuscripts concerned with arithmetic the sequencing and structural organization of content tended to be similar. Many of the topics were classified under names that were familiar to us because of our past interest in old mathematics textbooks—names like duodecimals, avoirdupois weight, troy weight, apothecaries' weight, reduction, rule of three direct, rule of three inverse, tare and tret, alligation medial and alternate, single and double fellowship, equation of payments, gauging, annuities, false position, and so on (for succinct summaries of the meanings of such terms, see Swetz, 1987, 2005). We knew that these names, and what they stood for, had arisen from commercial practices in the counting houses of European nations from about 1200 onwards (Swetz, 2005).

These cyphering books (our preferred term) usually comprised unlined sections of largish dimension (they were most often about 12.5″ by 8″) which had been sewn together. The ink was usually of a faded brown color, and headings often featured fine calligraphy. The manuscripts which dealt with navigation or surveying typically featured hand-drawn ink illustrations, some of which were filled-in with water colors. It was not unusual for someone who had originally prepared the cyphering books to have indicated, in writing, that the manuscript was "his book" or "her book," and the writer's name was often written on many pages of the same manuscript. Sometimes individual pages were numbered and dated.

We wondered why those who had prepared these manuscripts had been willing to devote so much time to them. After examining related literature, and searching the Internet for information on cyphering books, a coherent picture began to emerge. Then we happened to purchase the 671-page manuscript prepared by Charles Page at Christ's Hospital in London in the mid-1820s. This manuscript had the British royal crest on its front and back covers.

The high quality of the penmanship, calligraphy, and illustrations in Page's (1825) book stunned us. We were equally impressed with the tight, logical sequencing of the mathematics in the manuscript. We began to read as much as we could about the history of the Royal Mathematical School at Christ's Hospital. Among many things, we discovered that Isaac Newton had been among a distinguished group of advisers to the royal committee that had planned the Christ's Hospital mathematics curriculum and *modus operandi* (Allan, 1984; Howson, 1982). We learned that Newton's early mathematical training was possibly based on a cyphering book owned by Henry Stokes, a teacher at Grantham Grammar School (which Newton attended). Stokes's cyphering book dealt with the extraction of cube roots, surveying, mensuration, trigonometry, and geometrical constructions (Rickey, 1987). All of those topics featured in Charles Page's (1825) manuscript.

Samuel Pepys was the driving force behind the establishment of the Royal Mathematical School at Christ's Hospital in 1673 (Trollope, 1834) and given Pepys's (1995) own mathematical limitations, it was not surprising that he enlisted the help of a distinguished committee to help him develop and implement a mathematics curriculum and system of instruction at Christ's Hospital. That committee developed a program based on the cyphering tradition. We felt we wanted to find out more about the rationale for the cyphering tradition.

In our continuing search for the origins and essence of the cyphering tradition we came upon the Italian *trattati* or *libri d'abbaco* reckoning school practices that were so important in Europe during the period 1200–1600. The *trattati* literature taught us that around 1200 Leonardo Pisano ("Fibonacci") had played an important role in the introduction of Hindu-Arabic numerals into Europe and in the development of the *abbaco* tradition. We also learned that that tradition had played a large part in the development of European commercial practices, especially those concerned with reckoning (Franci, 2009; Spiesser, 2004).

Reading the research reports of Warren Van Egmond (1976, 1980) enabled us to understand more of the *abbaco* tradition, and the structure and genre of *abbaci* manuscripts. Van Egmond's (1976) dissertation helped us recognize that what we had previously regarded as "strange" business arithmetic topics (alligation, gauging, tare and tret, fellowship, rules of three, etc.) in fact represented part of the *abbaco* tradition. We conjectured that the cyphering tradition that we were investigating had emerged from the European *abbaco* tradition.

We now believe that that conjecture was correct—that, indeed, the cyphering books that we had located were artifacts generated toward the close of a 600-year tradition in mathematics education, a tradition that stretched back to a time well before printed arithmetics had been available. Furthermore, the fact that this tradition had continued to be important long after the emergence of printed books in the 15th and 16th centuries, forced us to reflect on what it was about the tradition that, some contemporaries believed, facilitated worthwhile learning.

In 2010, we became aware of the magnificent three-volume publication associated with the Michael of Rhodes' manuscript (Long et al., 2009). Michael was a 15th-century Genevan mariner, and the entire mathematical part of his manuscript was reproduced in one of the volumes. It was also translated into English. In the third volume, Franci (2009) analyzed the mathematics in Michael's manuscript. We were excited to find that so much of the mathematics in Michael's manuscript, written in the 1430s, was basically the same as that taught and learned via the cyphering approach in North America during the period 1607–1860.

Our own analyses of the cyphering books in our PDS led us to the conclusion that it was important that the early history of school mathematics education in North America be reconstructed. All the principal writers on that early history—Florian Cajori (1890, 1907), Walter S. Monroe (1917), Lao Genevra Simons (1924, 1936a, 1936b), Louis Karpinski (1925, 1926), David Eugene Smith and Jekuthiel Ginsburg (1934) and Patricia Cline Cohen (1982, 1993, 2003)—wrote at a time when the cyphering approach was no longer used, and had all but passed out of memory. With the exception of Simons, these historians focused largely on commercially-published textbooks that became available and were used by some teachers and learners during the 17th, 18th, and early 19th centuries. Even when some recognition was given to the importance of cyphering books, these manuscripts were presented as subordinate to the "printed books." Thus for example, Cohen (2003), in her article on "Numeracy in 19th-Century America," maintained that the cyphering approach "involved transcribing formulaic rules out of *printed textbooks* [our emphasis] word for word into the pages of individual copybooks, together with the exact examples offered by the text to illustrate each rule" (p. 44).

We gradually reached the conclusion that in fact the printed textbooks were much less used and, from a mathematics teaching and learning perspective, less historically important, than the cyphering books. The fundamental importance of cyphering books, and the theoretical bases behind the use of such books, seemed to have been overlooked by the historians. Simons was one historian who recognized the importance of cyphering books, but her analyses were largely restricted to cyphering books located within Harvard University's archives. She made it clear that she wished that she had been able to examine a larger number of "notebooks" (the term she used for cyphering books). She wrote:

> A great deal has already been said about the custom of keeping student notebooks during this period of difficulty in obtaining books from England and of printing books in the colonies. If all the notebooks now hoarded by descendants of graduates of the early American colleges or lying neglected and forgotten in attics and closets, if all these notebooks could be presented to the several college libraries or historical societies, the history of early American education would be greatly enriched. In these notebooks, there is found the content and scope of the curriculum of the day in evidence that is unmistakable. The number of notebooks located which bear the names of Harvard graduates is small. (Simons, 1936a, p. 588)

Other historians who have attributed at least some importance to data generated by cyphering books include L. W. B. Brockliss (1987), Patricia Cline Cohen (1982), Maria da Silva and Wagner Valente (2009) and Georges Glaeser (1984).

The need to gather evidence from a variety of sources is increasingly being recognized by historians of mathematics education. For example, Amy Ackerberg-Hastings (2009) pointed out that although textbooks can be valuable and interesting primary sources, of themselves they provide only limited insights into what went on in actual mathematics classes, or into what students learned. In order to develop a more coherent picture of early mathematics education practices, there is a need to synthesize insights gained from data gathered from multiple perspectives—including, but not confined to, textbooks, cyphering books, school and college timetabling and curriculum records, and written recollections by students.

We recognized the need to check our hypotheses against the largest possible set of cyphering books. Toward that end, we worked hard to expand the number of manuscripts in our own PDS. We also studied manuscripts in the Phillips Library in Salem, Massachusetts, the Houghton Library at Harvard University, the American University at Washington, DC, the Butler Library at Columbia University, the Clements Library at the University of Michigan, the Wilson Library at the University of North Carolina, and the Swem Library at the College of William and Mary, and the Rockefeller Library (both in Williamsburg, Virginia). We also examined cyphering books located in the New York Public Library, and in the London Metropolitan Archives and at Guildhall, London.

There are important "implemented curriculum" questions relating to the history of mathematics education in North America during the period 1607–1861 that cannot be adequately answered without examining data from a large number of cyphering books. Take, for example, the question of which algorithms were used for carrying out subtractions of natural numbers.

Suppose that around the year 1800, someone in New England wanted to find the value of 953 – 556. According to Smith (1925), there were four commonly-used algorithms, each of which had histories stretching back to at least the 16th century. The first of these was what Smith called the "complementary plan" (p. 98) by which a person would subtract 6 from 10, to get 4, and add this to 3, to get 7; then 1 would be "carried" and 6 would then be subtracted from 10, to get 4, and this added to 5, to get 9; then, after 1 had been carried, 6 would be subtracted from 9 to get 3. Thus, the answer would be 397.

Someone employing the second algorithm, which Smith (1925) called the "borrowing and repaying plan" (p. 99), started by subtracting 6 from 13 (10 has been added to the 3), to get 7; then the 10 would be repaid by adding 1 ten to 5 tens, to get 6 tens, and this would be subtracted from 15 tens (this time, 10 tens were added to the 5 tens), to get 9 tens; finally 6 hundreds would be subtracted from 9 hundreds, to get 3 hundreds, and so the answer would be 397. This method, often known as "equal additions," was recommended by Isaac Greenwood (1729), the first American-born author of an arithmetic designed for use in the American colonies.

The third algorithm was termed the "plan of simple borrowing" by Smith (1925, p. 100), but has subsequently become known as "decomposition." By this algorithm, the 953 is decomposed and is thought of as 8 hundreds, 14 tens, and 13 units, and from that are subtracted 5 hundreds, 5 tens, and 6 units, to give, once again, 397. This method was widely used in France in the eighteenth century (see, e.g., Barrème, 1747).

The other algorithm was called "making change" by Smith (2005, p. 101). By this method, one asks what do I have to add to 6 to get 13, and the answer is 7; then the person asks what do I have to add to 6 to get 15, and the answer is 9; and, finally, what do I have to add to 6 to get 9, and the answer is 3. So the overall answer is 397.

A quick check of the sections on subtraction in U.S. arithmetic textbooks available around 1800 indicated that Nicolas Pike (1788) recommended the use of the first of these four algorithms, Michael Walsh (1801) recommended the second, and Chauncey Lee (1797), Warren Colburn (1822), and Frederick Emerson (1832), the third. None of the textbooks used around 1800 that were consulted mentioned the fourth algorithm. As a whole, the textbooks threw little light on what was actually happening when students were trying to learn to subtract.

On consulting our Principal Data Set, we located 131 cyphering books that dealt with subtraction. In those, the algorithm (the "rule") being used could be identified unambiguously in only 50 cases. Of those 50, 32 used the second-named algorithm (the "borrowing and repaying" or "equal additions" plan), and 18 used the first-named algorithm (the "complementary plan"). None of them used either of the other two algorithms. The evidence gathered from the cyphering books indicated, clearly, that around 1800 one of the algorithms was the most favoured, and another algorithm, although less popular, was sometimes used. But no author of a cyphering book could be identified as having used the third of the algorithms ("decomposition")—which, incidentally, is the one most used in the United States in the 21st century. Also, no author could be identified as having used the fourth algorithm.

In short, although a textbook might have conveyed the author's *intended* curriculum by stating a rule, or showing a type example, what a student wrote in his or her cyphering book was much more likely to indicate the *implemented* curriculum (Meriwether, 1907). In any case, often neither the student nor his (or her) teacher owned a commercially-published textbook.

Cyphering books would continue to be widely used even when textbooks became more readily available. In the 17th century, Sir Roger L'Estrange had wondered "whether more mischief than advantage were not occasion'd to the Christian world by the invention of typography" (quoted in Briggs and Burke, 2010, p. 15), and it seems to us that even into the 19th century, many teachers, at all levels, remained suspicious of the value of textbooks, irrespective of whether they could afford to purchase them.

The Research Questions That Guided the Study

We then decided to formalize our research on the cyphering tradition, and to prepare this book setting out our findings and conclusions. That decision forced us to formalize our research by listing our research questions. We constructed the following five questions over a period of several months, and they have subsequently guided much of our research endeavour with respect to our analyses of the 212 manuscripts in our PDS.

1. Where and when were the cyphering books prepared, and by whom?
2. To what extent were the manuscripts consistent with the *abbaco tradition*, especially in relation to (a) mathematical content and sequencing; (b) genre; and (c) writing-as-also-arithmeticke considerations?
3. What theoretical base, if any, can be identified to encapsulate the educational purposes of the cyphering books?
4. Why, between 1840 and 1860, was there such a sharp decline in the use of the cyphering approach in the Unites States?
5. From an education learning perspective, what were the advantages and disadvantages of the cyphering approach?

We decided, too, that although those five questions would provide the foci for our research, we would not be constrained by them. That said, we will, nevertheless, answer each of the research questions in the final chapter of this book, and at that stage will comment on possibilities for further research into the cyphering tradition.

One last comment, here, might be of interest to readers. Most of the cyphering books in our PDS were purchased before 2009. During the period 2009–2011 we continued to check online and other auction lists for announcements of sales of cyphering books, that were originally prepared in North America, but fewer and fewer became available, and those that did tended to be of inferior quality. It seems that we made most of our PDS purchases at an opportune time, and that it would be difficult in the future for a collection of cyphering books similar in scope to our PDS to be established through purchases of individual manuscripts.

Each individual manuscript in the PDS was purchased independently of each other manuscript. Cyphering books in previously existing collections were not purchased and therefore do not form part of our PDS. We believe that that fact enhances the "representativeness" of our collection, and enables us to generalize more confidently than would have been possible if we had relied purely upon an analysis of cyphering book collections at any or all of Columbia University, Harvard University, the University of Michigan, the University of North Carolina, and the Phillips Library at Salem, Massachusetts. Of course, it was useful for us to examine these other cyphering book collections, for that provided us with the opportunity to check our emerging hypotheses, gained from studying the manuscripts in our PDS, against data generated by manuscripts in these other collections.

Chapter 5
The Principal Data Set

Abstract This chapter describes the authors' principal data set (PDS) for the study. This data set, which comprises 212 cyphering books, each of which was prepared between 1701 and 1861, is the largest single collection of cyphering books in the nation. In Chapter 5, questions associated with where and when the cyphering books were prepared, and by whom, are considered in the light of the author's analyses of their PDS. It is noted that the number of students preparing cyphering books fell sharply between 1840 and 1861, and the question is raised: Why did that happen?

The Extent of the Collection

Our handwritten manuscript collection comprises 293 manuscripts, most of which are cyphering books which originated somewhere in North America or in Europe. Of the 293 manuscripts, 212 are cyphering books prepared between 1701 and 1861 in those parts of North America which today make up the United States of America and Canada. It is this set of 212 North American cyphering books that has served as the principal data set ("PDS") for our analyses. As far as we know, our PDS is the largest single collection of North American cyphering books. Our two oldest cyphering books are from France and were prepared in 1695 and 1696. The two oldest North American manuscripts were prepared in 1701 and 1702. As far as we know, they are the oldest extant North American cyphering books.

The entries in most of the 293 handwritten manuscripts in our overall collection were written in English, but nine were written in Japanese, seven in German and four in French. Forty-three of our manuscripts *not* in the principal data set originated in England or Scotland. Only one of the 212 cyphering books in the PDS is written in a language other than English—it was written, around 1801, in "Pennsylvania Dutch," a German dialect (Parsons, 1976) often used in Pennsylvania. It is likely that many of the cyphering books in our PDS were not generated by apprentices in evening classes or by students who were being privately tutored or were attending so-called "writing schools," or "reckoning schools," or "subscription schools," in the American colonies (or, subsequently, in the United States of America).

Although we are not aware of the existence of an extant cyphering book prepared by someone living in any of the North America settlements during the 17th century, it is certain that such books would have been prepared. Karpinski (1925) cited original documents indicating that "cyphering" was part of the curriculum of a Boston "free school" in 1645, and at a school in the town of Dedham, in Massachusetts, in 1653. Also, in Boston, in 1682, because the Latin School was overcrowded, the town, after protracted discussions, voted to set up two schools "for the teaching of children to write and cypher" (Brayley, 1894, p. 31).

Excerpts from contemporary writing have convinced us that in the early North American settlements the cyphering approach was often used. Certainly, in the 18th and into the 19th century—until about 1840—the cyphering approach was common in the North American states and in Canada, especially (but not only) for the education of boys (Carlo, 2005). However, during the period 1840–1860 the use of cyphering books in North America dropped sharply, almost to the point of extinction. Only three manuscripts in our PDS were prepared after 1859, and the dates for each of those were 1860 or 1861. That raises a question which we shall discuss later in this book: why did such a longstanding tradition dissipate so quickly?

Analyses of the Principal Data Set

Details Relating to the Cyphering Books

Table 5.1 and Figure 5.1 summarize some details of the 212 manuscripts analyzed in the PDS. The 81 cyphering books in our collection *not* prepared by writers in North America have been excluded from analysis here. All of the manuscripts in the PDS were purchased by the authors, and we attempted to purchase any cyphering book that we knew was being sold either at an auction or by a private seller. Although, obviously, it would be foolish to claim that the sample of manuscripts was randomly selected from all manuscripts originally prepared, our sample nevertheless includes original manuscripts from *many* states, and it is almost certainly the only set of cyphering books which has that property.

Figure 5.2 shows, for each of the decades from 1780–1789 through 1870–1879, the ratio of the number of cyphering books in the PDS that were prepared in the United States of America to the number of millions of people living in the United States at the mid-point of each decade (for example, 1785, 1795, 1805, etc.). The population estimates were based on census returns.

Figure 5.2 provides a better picture than Figure 5.1 of the extent to which cyphering books were being prepared across the decades, because it takes into account the rapidly increasing population of the United States. Note that Canadian data were not taken into account in the calculations from which Figure 5.2 was obtained. Note, also, that Figure 5.2 does not attempt to control for the fact that cyphering books would have been lost over time.

In relation to Figures 5.1 and 5.2, we wish to point out that not only did we attempt to buy all North American cyphering books that we became aware were being offered at auctions during the period 2005–2011, but also, in over 90 percent of cases we succeeded in purchasing the manuscripts. Therefore, there is a sense in which our collection represents, albeit imperfectly, the set of North American cyphering books created during the period 1701–1861.

Table 5.1
Summary of the 212 North American Cyphering Books in the Principal Data Set

Period	Number of Manuscripts	Gender of Manuscript Writer		Locations where the Manuscripts were Prepared			
1700–1799	31	Male	23	Massachusetts	6	Canada	1
		Female	3	Connecticut	3		
		Unknown	5	Rhode Island	4	Pennsylvania	3
				New Hampshire	2		
				New York	5		
				New Jersey	2	Unknown	4
				Virginia	1		
1800–1839	147	Male	107	Massachusetts	28	Canada	3
		Female	24	Connecticut	10		
		Unknown	16	Rhode Island	12	Pennsylvania	24
				New Hampshire	7	Ohio	5
				Maine	6	Indiana	1
				Vermont	2	Michigan	1
				New York	17	North Carolina	1
				New Jersey	4	Tennessee	1
				Maryland	7	Alabama	1
				Virginia	1	Missouri	1
						Unknown	15
1840–1861	34	Male	24	Massachusetts	5		
		Female	8	Connecticut	–		
		Unknown	2	Rhode Island	–	Pennsylvania	12
				New Hampshire	1		
				Maine	2	Ohio	–
				Vermont	–	Indiana	–
				New York	3	North Carolina	1
				New Jersey	1	Arkansas	1
				Maryland	4		
				Virginia	2	Unknown	2
Overall 1702–1861	212	Male	154	Massachusetts	39	Canada	4
		Female	35	Connecticut	13		
		Unknown	23	Rhode Island	16	Pennsylvania	39
				N. Hampshire	10	Ohio	5
				Maine	8	Indiana	1
				Vermont	2	Michigan	1
				New York	25	Missouri	1
				New Jersey	7	Tennessee	1
						Arkansas	1
				Maryland	11	North Carolina	2
				Virginia	4	Alabama	1
						Unknown	21

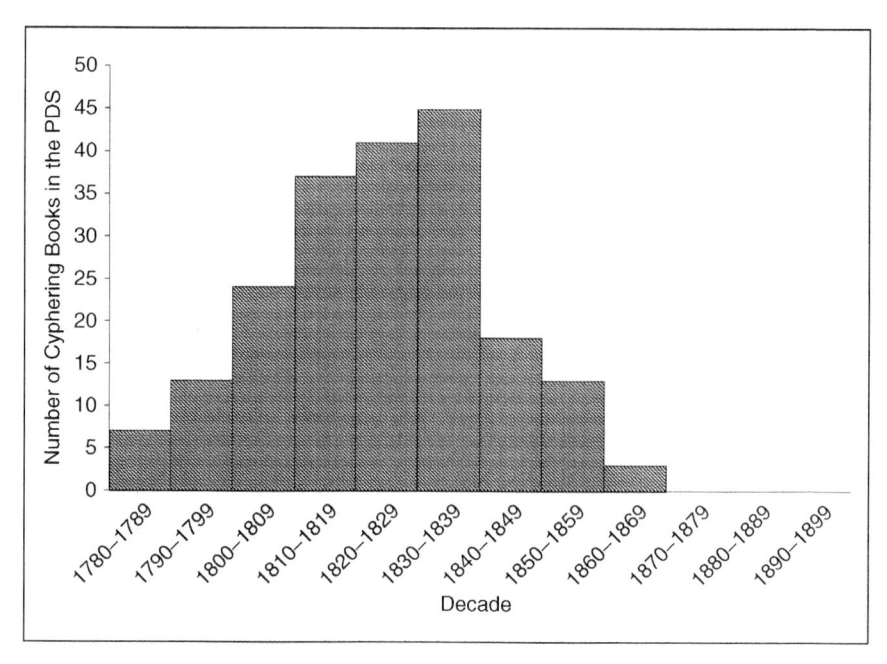

Figure 5.1. Number of cyphering books per decade (1780–1789 through 1870–1879) in the PDS. (Note that pre-1780 cyphering books in the PDS were not included in this analysis.)

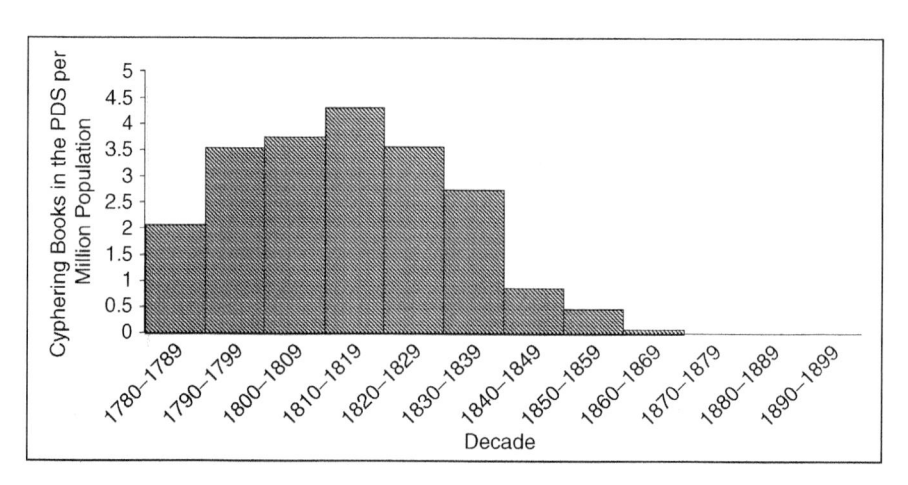

Figure 5.2. Ratio of cyphering books prepared to millions in the US population per decade, 1780–1789 through 1870–1879, in the PDS. (Canadian data excluded from the analysis)

We believe that the other significant collections of cyphering books are not nearly as representative as ours. Our PDS is, we believe, the only collection of cyphering books that originated from more than 10 of the colonies/states which existed during the period 1700–1861.

Thus, for example, most of the cyphering books held in the Phillips Library in Salem, Massachusetts, and many held in the Houghton Library at Harvard University, originated in New England states. Many cyphering books in the Butler Library at Columbia University originated from families living around New York, and cyphering books held by the Huguenot Historical Society were prepared by children from families with identifiable religious and cultural similarities. Most of the cyphering books in the Southern Historical Collection in the Wilson Library at the North Carolina University at Chapel Hill and the Swem Library at the College of William and Mary, and the Rockefeller Library, Williamsburg, originated from North Carolina (Coon, 1915; Doar, 2006) or Virginia. Many of the cyphering books held by the Clements Library at the University of Michigan were prepared by persons connected in some way with the Revolutionary War of the 1770s and 1780s.

Although many PDS cyphering books were prepared by persons who attended educational institutions such as schools, colleges, and private academies, in most cases we do not know the name of the educational institutions involved. It was not unusual for a cyphering book to become a family possession, to be passed on, and used by others. In such a case, different names were likely to appear in different parts of a manuscript.

We suspect, but cannot prove, that many who prepared cyphering books in our PDS were apprentices taking special evening classes. Some were prepared by midshipmen who were required to take classes during long voyages (Morrice, 1801). For these apprentices, and midshipmen, a cyphering book was likely to have been the only book, other than a Bible or an almanac, that they ever owned. In such circumstances, it is understandable that those who prepared the manuscript thought of it as a special possession in which they might write, proudly: "This is my book." There was also a competitive context, in which a student would have wanted his cyphering book to be better than anyone else's.

Cyphering books prepared in local public or subscription schools usually dealt mainly with elementary arithmetic—including numeration, notation, the four operations, compound operations, reduction, practice, vulgar and decimal fractions, and one or more of the rules of three. In the context of daily school activities, although cyphering books prepared in public schools were important they were not as personal as cyphering books prepared by apprentices.

Some cyphering books were prepared by college students and, we conjecture, these also tended to be less elaborate than those prepared by apprentices. Students attending colleges like Harvard, Yale and Columbia were immersed in a curriculum in which the chief emphases were the classics—especially Latin—and rhetoric (Rudolph, 1977). Some of these students loved mathematics, but at college it was not intended to be a major focus.

Issues associated with the circumstances surrounding those who prepared cyphering books remain for future researchers to identify and investigate.

The Three Periods

1700–1799. This period was deliberately chosen so that all 18th-century manuscripts in our collection would be considered within the one category. We are confident that the fifty-three 18th-century cyphering books in the Phillips Library, the forty 18th-century manuscripts held by the Houghton Library at Harvard University, and our own collection of thirty-one 18th-century manuscripts represent the three largest collections of 18th-century North American cyphering books.

1800–1839. For most of this period, education practices responded to local influences largely within a *laissez-faire* education environment. There were no district or college inspectors, no externally-set written examinations, and hardly any teachers who had been formally trained for teaching. From the 1820s onwards, however, systems of district high schools slowly began to emerge, and this resulted in mathematics curricula gradually becoming more standardized. Our set of 147 US cyphering books for this period (in our PDS) and the set of cyphering books for the same period held in the Phillips Library, Salem, are easily the largest sets of cyphering books in the United States for the period.

1840–1861. The set of 34 US cyphering books for this period in our PDS is the largest set of extant cyphering books for the period in the United States. During this period the cyphering approach to arithmetic became non-viable as a result of rapidly changing education environments characterized by increasing external control over curricula, externally-set examinations, and new pedagogical practices associated with the establishment of normal schools and the moves towards graded schools and high schools. The numbers of students in schools increased rapidly, so that in cities and in larger towns it became possible to place students in classes according to age, or gender, or, as was sometimes the case, perceived intellectual ability. In rural areas, one-teacher schools were still common, however, and in these it was more likely that the cyphering approach would continue to have been used.

Entries in the second column of Table 5.1, and in Figures 5.1 and 5.2, suggest that the number of PDS manuscripts originating during the period 1840–1859 was significantly less than for the 20-year period 1820–1839. Figures 5.1 and 5.2 also suggest that cyphering books prepared in North America after 1861 are hard to find and, indeed, we are not aware of the existence of any such manuscripts.

On locating and analyzing about 440 cyphering books that were *not* part of our PDS, and were held in various places across the United States, we found that about 100 of these had been prepared in the 18th century (the earliest being dated 1711), but none had been prepared after the 1850s. These additional analyses supported our thesis that the cyphering tradition was alive and well in the United States in the 1820s and 1830s, but the period 1840–1865 witnessed its demise. Such a sudden change warrants the attention of historians.

Analyses of the Principal Data Set

The Gender of Writers

The third column of Table 5.1 indicates that almost 19 percent of the cyphering books in the PDS for which the gender of the writer was known, were prepared by females. That statistic is in line with an analysis by Halwas (1997), which showed that about 17 percent of the original owners of 208 US mathematics textbooks published between 1760 and 1850 were females.

After examining cyphering books, mainly located in the Phillips Library, Salem, and Harvard University Library, Middlekauff (1963) argued that although some 18th-century New England females did prepare cyphering books, the topics they studied involved only elementary rules and computations. Cohen (1982) stated that "the Harvard collection of cyphering books contains six by colonial girls, but none went beyond problems in simple interest" (p. 251n). E. I. Monaghan (2007) maintained that girls who cyphered did not proceed beyond the rule of three.

However, of the three 18th-century cyphering books in our collection prepared by females, two went beyond simple interest and the rule of three. A page from the 1767 cyphering book prepared by Sally Halsey (Appendix A, m/s #4) has been reproduced in Figure 5.3. Lucy Starr's (1797) cyphering book (Appendix A, m/s #29) also contained higher-level arithmetic.

We know that some 17th-century girls in England prepared cyphering books (Smith, 2005), and there can be no doubt that many schoolgirls in 18th-century and early 19th-century North America created cyphering books. In New Orleans in the 1830s, for example, girls who attended the Ursulines Academy followed a common curriculum that included elementary arithmetic as well as reading, writing, grammar, history, geography, plain sewing and knitting (Doar, 2006). In the 1790s, all 365 girls who attended the Moravian Seminary for Young Ladies in Bethlehem, Northampton County, Pennsylvania (Moravian Publication Office, 1876) prepared cyphering books. The contents of a cyphering book begun in December 1790 by Susan Shimer, a student at the school, demonstrate that the arithmetic was not always elementary. All of the pages of this cyphering book can be seen on the web-page of the Bethlehem Digital History Project (2009).

The arithmetical content of the beautiful cyphering book prepared in North Carolina around 1780 by sisters Martha and Elisabeth Ryan was non-trivial, with the sisters dealing with topics that went well beyond the rule of three in complexity (Doar, 2006).

The relatively advanced arithmetical content in the cyphering books of Susan Shimer and the Ryan sisters notwithstanding, our analyses of manuscripts in our PDS supported Cohen's (1982) and Middlekauff's (1963) contention that during the 18th and early 19th centuries, girls rarely studied mathematics beyond arithmetic. Only two of our 22 "non-arithmetic manuscripts" were prepared by females. One dealt with elementary surveying, and was prepared by Betsy King (Appendix A, m/s #193) in the 1840s. The other was concerned with elementary algebra, and was prepared by Sarah Pierce in 1857 (Appendix A, m/s #207). Note, however, that we have seen two counter-examples to the "elementary-arithmetic-for females" thesis. The Houghton Library, at Harvard University, holds a 1797 manuscript by Phoebe

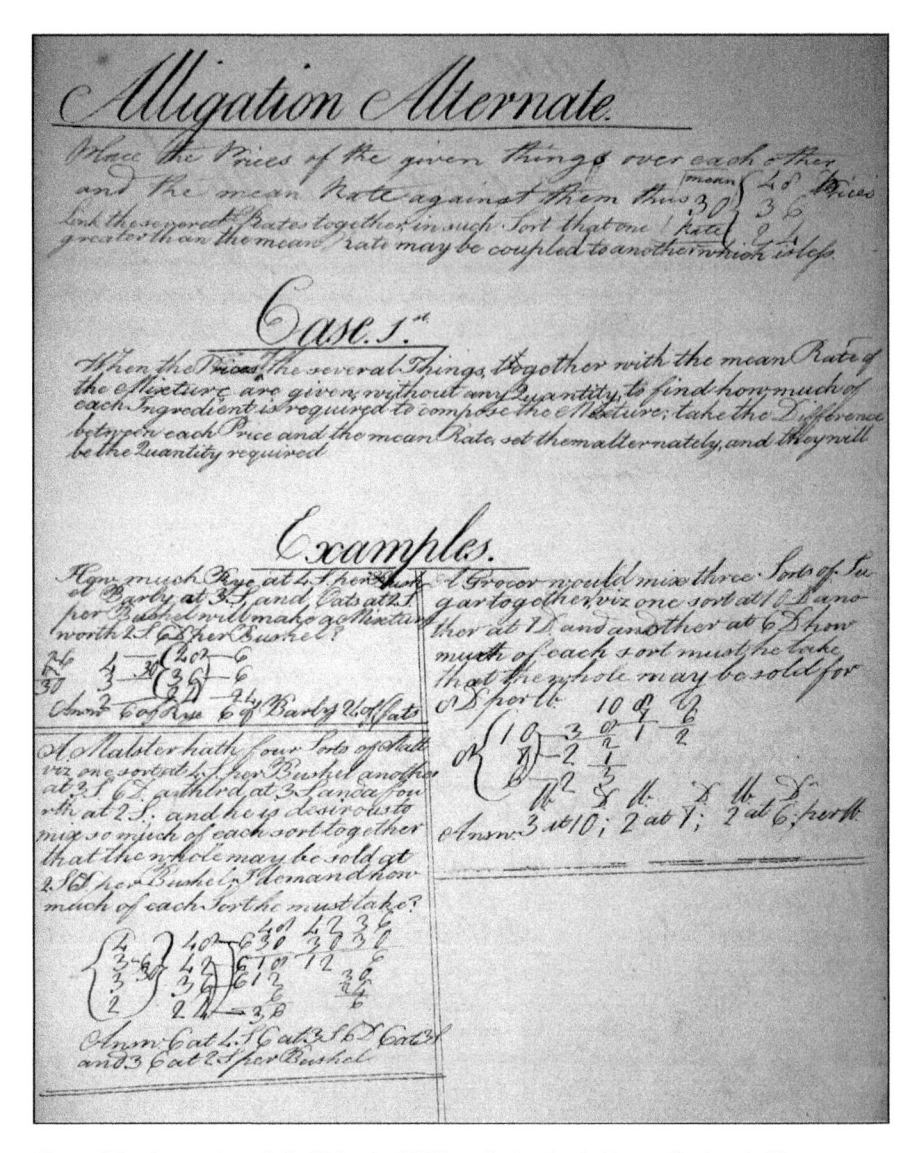

Figure 5.3. A page from Sally Halsey's (1767) cyphering book (Appendix A, m/s #4).

Folger which was concerned with algebra to quadratic equations. The same library also holds a 1792–1793 manuscript by Sally (Unknown family name), which was concerned with trigonometry and surveying. Phoebe Folger was the sister of Walter Folger, who was a well-known mathematician and clockmaker.

Locations of Manuscripts

Of the 191 manuscripts in the PDS for which locations of writers could be determined, 170 were prepared in a New England state, or in Pennsylvania, New York, New Jersey or Maryland. Only 17 manuscripts derived from other states—5 from Ohio, 4 from Virginia, 2 from North Carolina, and one from each of Alabama, Arkansas, Indiana, Michigan, Missouri and Tennessee—and 4 manuscripts were prepared in Canada. No manuscript in the sample was prepared by a writer in any of California, Delaware, Florida, Georgia, Illinois, Iowa, Kentucky, Louisiana, Minnesota, Mississippi, Nevada, Oregon, South Carolina, Wisconsin, or Texas—all of which had been proclaimed states before 1861. The longer history of cyphering books in Continental Europe than in Great Britain is, perhaps, reflected in the fact that more than one-third of the non-Canadian cyphering books for which the locations are known were originally prepared in New York, New Jersey or Pennsylvania—colonies/states that were often preferred by settlers from continental Europe, but were less likely to be favored by settlers from Great Britain (Burke & Burke, 1758).

Why were there so few cyphering books in the collection from the Southern and mid-Western settlements, colonies or states? The main contributing factor was probably the tradition within many states that only a small number of people needed a formal mathematical education beyond the stage of counting and numeral recognition. As Cohen (2003) has stated, an Anglo-American assumption prevailed until well into the 1800s "that numeracy was a business skill needed chiefly by boys heading for careers in business and trade" (p. 45).

In the mercantile and maritime cities of Boston, New York, Philadelphia, Salem, Newport and Providence, it made some sense for children to want to become business clerks, or reckoners, or navigators, or surveyors. In most, but not all other places, parents tended to believe that they needed as much assistance in their daily enterprises as they could get from their children in order that their families would be able to survive. Furthermore, parents who wanted their children to proceed to higher studies realized that even if qualified tutors could be found, payments needed for tuition in cyphering might be prohibitive. That was especially true of families in isolated rural communities.

When we were examining cyphering books held by the University of North Carolina we were struck by the fact that students in the South often prepared books throughout the whole year. We conjecture that plantation owners had slaves to meet their labor needs, and therefore their children were free to prepare cyphering books at any time in the year.

Only four cyphering books in the PDS were prepared in Canada, but that should not be regarded as evidence that the cyphering tradition was not important in Canada. In 1809, Jean Antoine Bouthillier stated, in the preface of what appears to have been the first mathematics textbook printed in Canada, that Canadian teachers had long been obliged to have students copy lengthy principles and rules of arithmetic into their "copy books," and he hoped that his book would "remedy that inconvenience" (quoted in Michalowicz & Howard, 2003, p. 85).

The Mathematical Content in the Cyphering Books

Of the 212 cyphering books in the PDS, 191 dealt primarily with arithmetic, 7 with algebra, 8 with theoretical geometry and/or trigonometry, 3 with navigation and 3 with surveying. This probably reflected the situation in Great Britain, for the Duke of Newcastle's (1858–1861) government commission that investigated elementary education in 1824 public weekday schools in England reported that although almost 70 percent of the students studied arithmetic, only about 2 percent studied any of algebra, Euclid or mechanics. In the private schools in England, only 34 percent studied arithmetic, and about 4 percent studied any of the other branches of mathematics (see Howson, 1982; Rogers, 1999).

Arithmetic. Evidence from the PDS strongly suggests that throughout the period 1701–1861 the dominant content emphasis for North American students learning mathematics was arithmetic. The sequences of arithmetic topics in the cyphering books were fairly consistent with sequences found in the *abbaci* and in popular commercially-published texts prepared by authors like Edmund Wingate (1630), Edward Cocker (1720, 1725), Thomas Dilworth (1797), James Hodder (1661), William Leybourn (1690), Zachariah Jess (1811) and Nicolas Pike (1788). Although it is not possible to determine whether any or all of these authors were conscious of the over-riding influence of the *abbaco* tradition on their thinking as they prepared their texts, the large extent of the influence cannot be denied.

Even when textbooks provided commentary on more modern topics, there could be no guarantee that topics in textbooks were the same as those studied in classrooms—that was true even if an instructor had written the textbook. Amy Ackerberg-Hastings (2009), for example, has shown that in the early 1800s when the eminent Scottish mathematician/philosopher, John Playfair, was teaching courses in which he himself had written the set textbook, the notes taken by his students did not always follow the textbook. Textbooks might have taken account of advanced algebra or calculus, but there was no guarantee that Playfair expected his students to know, or to learn, such material. In other words, the content of the handwritten notes prepared by students often differed from what was in the textbook (Ackerberg-Hastings, 2009).

Such a finding raises the issue whether 18th- and 19th-century textbooks influenced mathematics teaching and learning as much as many historians have supposed. We hold a handwritten manuscript prepared by a University of Aberdeen mathematics student (a certain "William Swewnight") of the early 1860s, in which it is clearly stated that the notes were "used" by four other students—and there is the suggestion that these other students had not attended the lectures on which the notes were based. In this case, notes based on a lecture (perhaps dictated at a "lecture") seemed to have become a pseudo-textbook. According to Bellhouse (2010), there is evidence that it was common for students to copy from handwritten manuscripts prepared by others, rather than take notes at lectures or copy directly from published books. Still, since the proprietors of some of the dissenting academies in England in the 18th century—like, for example, Alexander Ewing (1799), Francis Walkingame and John Bonnycastle—were authors of published textbooks, it is likely that in their schools, at least, cyphering was largely based on commercially-prepared textbooks (Cockburn et al., 1969).

Students at Harvard and Yale Colleges in the 18th-century (and, in the case of Harvard, perhaps before that, in the 17th century) often purchased a mathematics textbook (e.g., Ward, 1719). As the 18th century progressed, that was also true of students at other colleges (Cajori, 1890). But, textbook considerations notwithstanding, it was standard for students to prepare cyphering books based on handwritten texts owned by the instructors. For example, there is evidence that at Harvard in the 1730s, students copied entries from a text owned by Isaac Greenwood, the first Hollis Professor of Mathematics (Simons, 1924). It was these notes, rather than any notes in any textbook, or any notes given in lectures, that provided the foci for the regular recitation sessions.

This practice continued into the 19th century. Fred Rickey and Amy Shell-Gellasch (2010) have related how Abraham Wendell created his own cyphering book at West Point in 1814. The problems were taken from Hutton's (1812) *Course of Mathematics*. According to Rickey and Shell-Gellasch, graduating students at West Point would lend their recently-completed cyphering books to new cadets. Rickey and Shell-Gellasch provided the following delightful quotation from a Professor Albert Church (West Point Class of 1828) who, in 1864, recalled his cyphering-book experiences at West Point:

> Of course, the real teachers in these subjects were those cadets who made careful notes, finished their drawings early in the day, made the demonstrations to their classmates, and lent their drawings for copying. Great skill was required in making these copies. A clear and large pane of glass placed on the top of the washstand, a lighted candle underneath the finished drawing on the glass, and the paper for the copy on the top, and every point quickly marked with a pencil. (p. 394)

Entries in the Wendell and the Griswold notebooks can be precisely matched with Hutton's text (Rickey & Shell-Gellasch, 2010). Our point is, however, that those responsible for mathematics education at West Point made sure that every cadet carefully prepared a cyphering book.

In fact, until around 1850, wherever and whenever arithmetic was taught in colleges and schools the *abbaco* tradition defined the content, scope and, despite Cohen's (2003) assertion to the contrary, the sequence of the curriculum. The level of consistency was sufficiently great that colleges specified the amount of arithmetic required to be known by applicants for admission by phrases like "arithmetic to the rule of three" (at Princeton), and "arithmetic to the rule of reduction" (at King's College, later Columbia University) (Broome, 1903). Of the 191 cyphering books dedicated to arithmetic in the PDS, 39 dealt with elementary arithmetic only (they did not go beyond "the rule of three direct"), 140 were concerned with middle-level as well as with elementary topics, and only 12 generally dealt with more advanced topics.

At the elementary level, the first topics likely to be covered were notation and numeration, followed by the four operations on whole numbers. Then would come a selection from the following: compound operations, reduction (including measurements), money and currency conversion, reduction, vulgar (or "common") fractions, decimals, percentage, simple interest, loss and gain, the single rule of three, and various elementary business arithmetic topics,. This sequence of topics was consistent with that in a document (reproduced in Karpinski, 1980, p. 85) written in 1786 by a school committee of the town of Providence, in Rhode Island. The placement of vulgar and decimal fractions might vary, but otherwise the order of topics would be constant. Sometimes, only a beginning would be made: for example, Walter Monroe (1917) cited a document which indicated that in New York in 1815, "of the children studying arithmetic, 208 were in addition and subtraction, 110 in multiplication and division, 15 in compound numbers, and only 10 in reduction and the rule of three" (p. 41).

In Lucy Starr's (1797) cyphering book (Appendix A, m/s #29), the order of topics was simple subtraction, multiplication, division, Federal money, compound addition, subtraction, multiplication, division, reduction ascending, reduction descending, practice, and rule of three direct. The use of the adjective "compound" in that list implied that quantities (money, weights and measures, etc.) were to be added, subtracted, multiplied or divided. The units for different quantities—which included dry measure, liquid measure, wine measure, cloth measure, apothecaries' weight, avoirdupois weight, troy weight, square measure, cubic measure, and time—needed to be known (see Appendix C).

The rule of three direct was traditionally regarded as the final step for those wishing to complete a course in elementary arithmetic. For those who went further, to what we have termed middle-level arithmetic, the emphasis on business arithmetic progressively increased. The only middle-level topics not directly concerned with business arithmetic were lowest common multiple, greatest common divisor and duodecimals, but these topics were not to be found in most cyphering books.

Thus, for example, Lucy Starr (1797) (Appendix A, m/s #29) followed her elementary work with middle-level content on the single rule of three inverse, discount, equation of payments, and fellowship. Both John Anderson (1785) (Appendix A, m/s #14) and Sally Halsey (1767) (Appendix A, m/s #4) solved problems related to all of those topics, before proceeding to compound interest, rebate, barter, loss and gain, exchange, compound fellowship, double rule of three, alligation medial, alligation alternate, alligation partial, alligation total, tare and tret, and single and double position.

"Advanced" level topics included involution, evolution, annuities, gauging, permutations and combinations, arithmetical and geometrical progressions, surds and logarithms. Problems associated with most of those topics were presented and solved in PDS cyphering books prepared by Hugh Ross (1797, Appendix A, m/s #30), Warner Green (c. 1800, Appendix A, m/s #34), Sarah Mead (1806, Appendix A, m/s #46), Isaac Heaton (1809–1813, Appendix A, m/s # 55), Leonard Levi (c. 1810, Appendix A, m/s #57), and a few other students.

Most cyphering books provided brief introductions to the main topics. Because these were cyphering books, rules were stated, cases considered, type problems presented and solved, and exercises stated and solved (see Figure 5.4, which shows

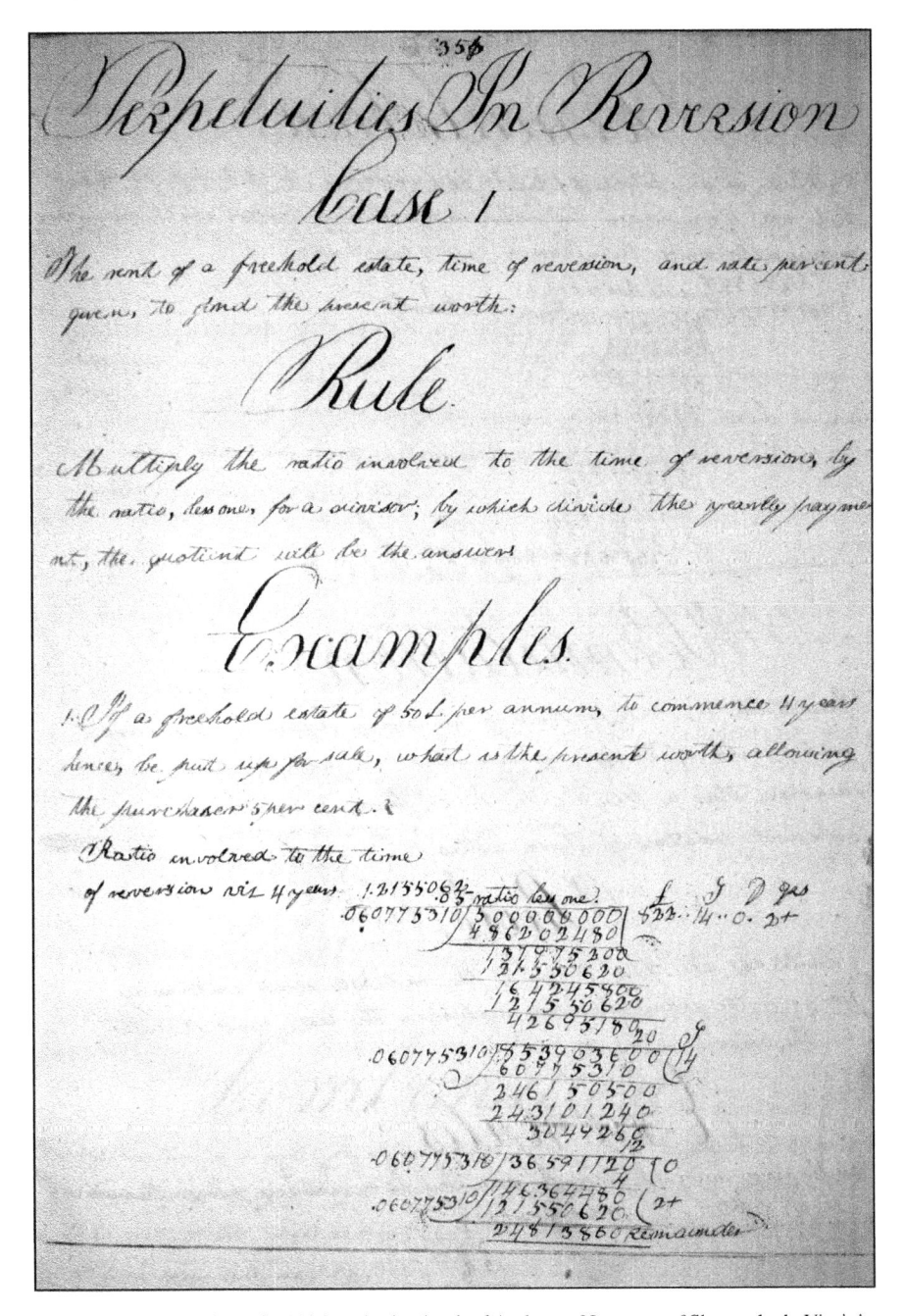

Figure 5.4. A page from the 1824 cyphering book of Anderson Newman, of Shenandoah, Virginia.

a page from the cyphering book of Anderson Newman of Virginia (1824, Appendix A, m/s #112). Reasoned statements, indicating where rules came from, or why they worked, were rarely given. In most cases, "solutions" consisted of nothing more than numerical calculations. These were *never* marked as correct or graded—presumably because teachers and tutors would not permit their students to enter statements into their book unless the correctness of the work had been established beforehand. Only 7 of the 191 manuscripts that focused on arithmetic included *recreational* problems.

Algebra, Geometry, Trigonometry, Surveying and Navigation

Algebra. Only seven cyphering books in the PDS focused on algebra (and another two dealt with elementary algebra after first having been concerned with arithmetic). Those data are consistent with Jeremy Kilpatrick and Andrew Izsák's (2008) and Lao Genevra Simons' (1924) claims that algebra was hardly studied in 18th-century schools in North American colonies and settlements. Nevertheless, given that algebra was taught in most colleges during the first half of the 19th century and, from the 1820s, in the high schools (Kilpatrick & Izsák, 2008), we were surprised that so few of our 19th-century PDS cyphering books had a focus on algebra.

Our investigations revealed that it was not until 1838 that the number of high schools in the United States of America totalled 10, and even in the 1870s, only a tiny proportion of American youth attended a public high school (Bureau of Education, 1905). From a national perspective, although algebra was part of the curricula of the early high schools established in New England states in the 1820s and 1830s, this did not bring a large increase in the number of students taking algebra. However, when the number of high schools, and normal schools, increased rapidly after 1840 (Bureau of Education, 1905), algebra began to be more widely studied at the secondary school level. That was related to the fact that Harvard, Columbia, Yale and Princeton had decided, in 1820, 1821, 1846, and 1848, respectively, that students wishing to be admitted had to demonstrate a knowledge of elementary algebra (Overn, 1937).

The year 1814, marked the publication of the first dedicated algebra text written by a North American author. That author was Jeremiah Day (1814), Professor of Mathematics at Yale College (Ackerberg-Hastings, 2000; Simons, 1936b). Elementary algebra had been studied by some college students in the 18th century (Hornberger, 1968; Kilpatrick & Izsák, 2008; Rudolph, 1977; Simons, 1924), but very little algebra—often no algebra—appears to have been part of the pre-college curricula followed in colonial schools.

An analysis of the content of the seven algebra manuscripts in our PDS revealed a fairly stable sequence in which elementary algebra problems were presented and solved. First came notation and elementary rules for the four operations; then followed factoring, algebraic fractions, linear equations and associated word problems. Those who progressed beyond the elementary level studied quadratic equations, arithmetical and geometrical progressions, the binomial theorem, and logarithms. The algebra books emphasized a "rules" and "cases" approach, with model examples and their solutions being followed by exercises. Explanations were kept to a minimum, and usually the standard of penmanship and calligraphy was high. The concept of a function was unknown in the schools, and graphs were not to be seen in the cyphering books or in school textbooks (e.g., Sangster, 1865).

Geometry. Of the 212 handwritten manuscripts in the PDS, only eight had a primary focus on geometry or trigonometry. Each of these presented a Euclid-like axiomatic approach to geometry, and sometimes elementary trigonometry was considered. One might be tempted to conclude that in North America before 1860, formal geometry and trigonometry, like algebra, was rarely studied, except at college level. However, that would be an over-simplification, because those preparing to be surveyors or navigators typically dealt with definitions of geometrical figures and straight-edge/compass constructions, as well as trigonometry (often spherical as well as plane), and logarithms.

Surveying and navigation. Three PDS manuscripts were dedicated to surveying and three to navigation. Given the obvious need in the colonies for navigators and surveyors (Kiely, 1947), one might have expected more related manuscripts in our PDS. However, the level of mathematics demanded of someone wishing to become an efficient navigator or surveyor was considerable, and although there were private tutors in the big cities claiming to be capable of teaching those subjects (see, e.g., Karpinski, 1980; Kiely, 1947; Seybolt, 1935), one wonders how many of them could have coped with what was needed to bring students to the stage where they could deal with the mathematics that was required. The following account, from a British text published in 1868, throws light on the associated educational issue:

> When a boy is too disobedient to be governed at home, too inattentive to learn at school, and too idle to work at "a place," he is then qualified for sea. He, perhaps, learnt while at sea that a knowledge of navigation would be useful, and he resolves to redeem 12 or 13 lost years of his life by the desperate efforts of a month. He betakes himself to the Mechanics Institution, and something like the following dialogue takes place in the mathematical department:
>
> *Teacher:* What do you wish to learn?
> *Sailor:* Double altitudes and lunars.
> *Teacher:* Do you understand trigonometry?
> *Sailor:* No!
> *Teacher:* Do you know anything of geometry?
> *Sailor:* No!
> *Teacher:* Do you understand decimals?
> *Sailor:* No!
> *Teacher:* What did you learn when you went to school?
> *Sailor:* I think I went as far as multiplication.

(*Chambers' Information for the People,* 1868. Vol. 1, pp. 733–734)

The point is that the mathematical expertise and practical experience that students needed in order to respond adequately to the curricular demands of any decent course in navigation or surveying were much more than would have been provided by a standard course in arithmetic.

If "curricular importance" is measured by the number of enrolled students, then Edmond Kiely's (1947) claim that "surveying formed an important part of the

American mathematical education program during its first two centuries" (p. 245), is misleading. Given the need for surveyors in early colonial society, one might have expected the subject to be studied, formally, by more students than appears to have been the case. The hard fact was that in order to become a surveyor a student needed more prerequisite mathematical skills than those provided by the pragmatic and rushed forms of mathematics education that early colonial societies provided.

Trigonometry was rarely taught in one-roomed schoolhouses because relatively few teachers had had any experience with it, and very few students were ready to learn it, anyway. In most Latin schools, the classics held sway, and there was no inclination, or time, for practical subjects. Apprentices who attended evening classes had rarely studied trigonometry, and therefore were unlikely to be in a position to learn the theoretical content, and to acquire the skills, needed to become effective practicing surveyors.

The 14 manuscripts in the PDS that were primarily concerned with geometry, trigonometry, surveying or navigation had a more descriptive genre. There were still plenty of rules, cases, type problems, and exercises, but it was unlikely that many of the students could have solved most of the exercises without a lot of help. There were exceptions, however—the manuscript prepared by William Turnbull (1780, Appendix A, m/s #13) of New Jersey, was "Christ's Hospital-like." Figure 5.5 shows a page from Turnbull's book.

The end of an era in curriculum. Of the 212 manuscripts in the PDS, 15 dealt, primarily, with algebra or geometry (7 for algebra, 8 for geometry). Those statistics attest to the end of one mathematics curriculum epoch and the beginning of another, for according to Donoghue (2003a), in the 20th century, algebra and geometry would be the subjects "the most common to the experience of secondary school students across the century and across the country" (p. 329).

Genre Considerations

The standard definitions-rules-cases-type examples-exercises approach which reflected the *abbaco* heritage (Heeffer, 2008; Van Egmond, 1976), were evident throughout the cyphering books in the PDS. However, the early *abbaci's* characteristic rhetorical "dialogue" genre for the statement and solution of problems (Van Egmond, 1980), by which a problem poser asked a problem solver a series of leading questions, and then answered them, was only occasionally found in the cyphering books that made up the PDS.

In the 18th- and 19th-century manuscripts, the dialogical *abbacus* genre was replaced by a more compact genre in which the statement of a type problem was followed by calculations that produced the correct answer. Then a related exercise would be given, followed by calculations that somehow generated an answer (which was often marked "Ans"). Figure 5.6, reproduced from Thomas Prust's (1702) cyphering book (Appendix A, m/s #2), shows two "type" examples for "substraction"—an alternative spelling to "subtraction"—and associated solutions. Thomas Prust's declaration of ownership (see Figure 5.6) was common among those who prepared cyphering books. The manuscripts were their books, to

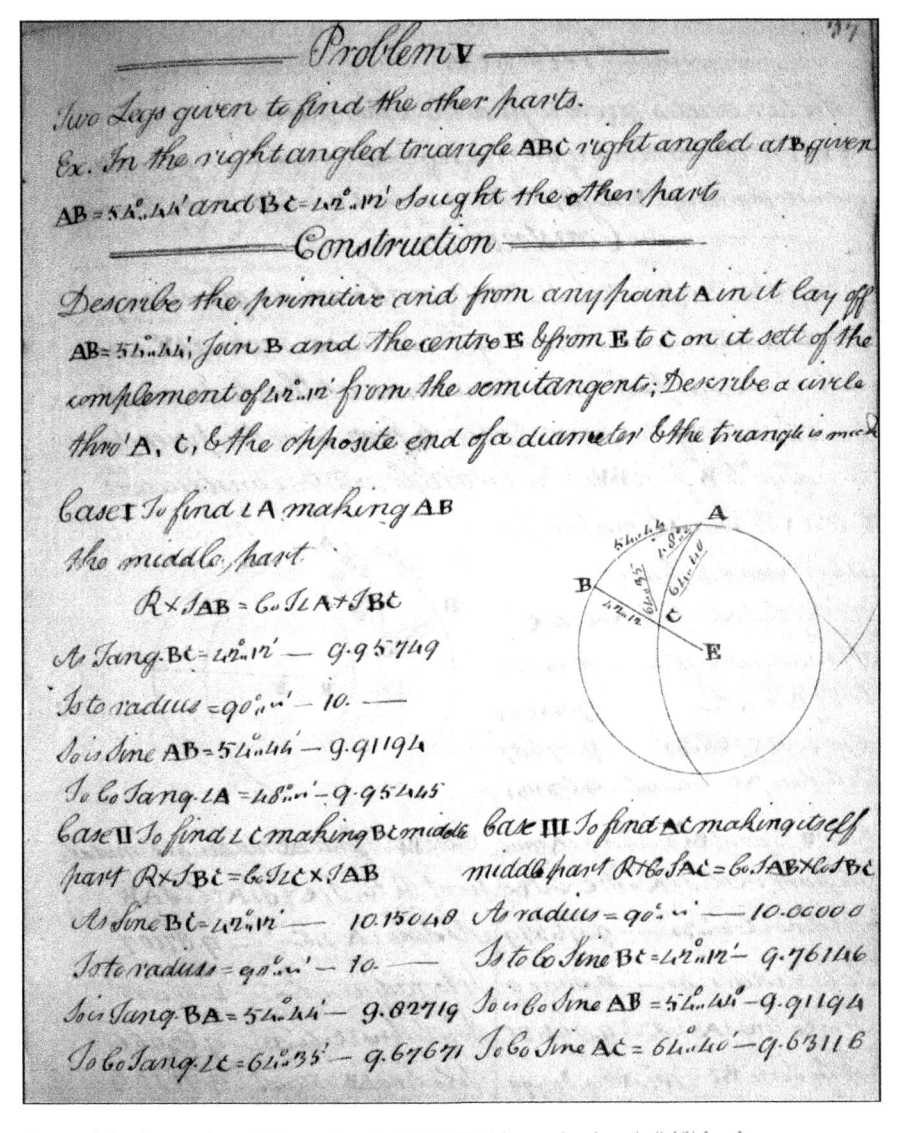

Figure 5.5. A page from William Turnbull's (1780) (Appendix A, m/s # 13) book.

guide them for life. For centuries, pride of ownership had been a part of the cyphering tradition in Europe—it is to be found, for example, in Michael of Rhodes's 15th-century book (Rossi, 2009).

Of the 212 cyphering books only 27 had at least one diagram. Written solutions rarely incorporated equal signs or logical connectives. Most students solved problems as quickly and as briefly as possible. Such brevity was regarded as desirable (French, 1869).

Our analyses of the content of our cyphering books pointed to the conclusion that a compact genre was expected of arithmetic students during the 160-year period covered by the PDS. Figure 5.7, from the cyphering book prepared by an anonymous student around 1815 (Appendix A, m/s #79), illustrates the genre.

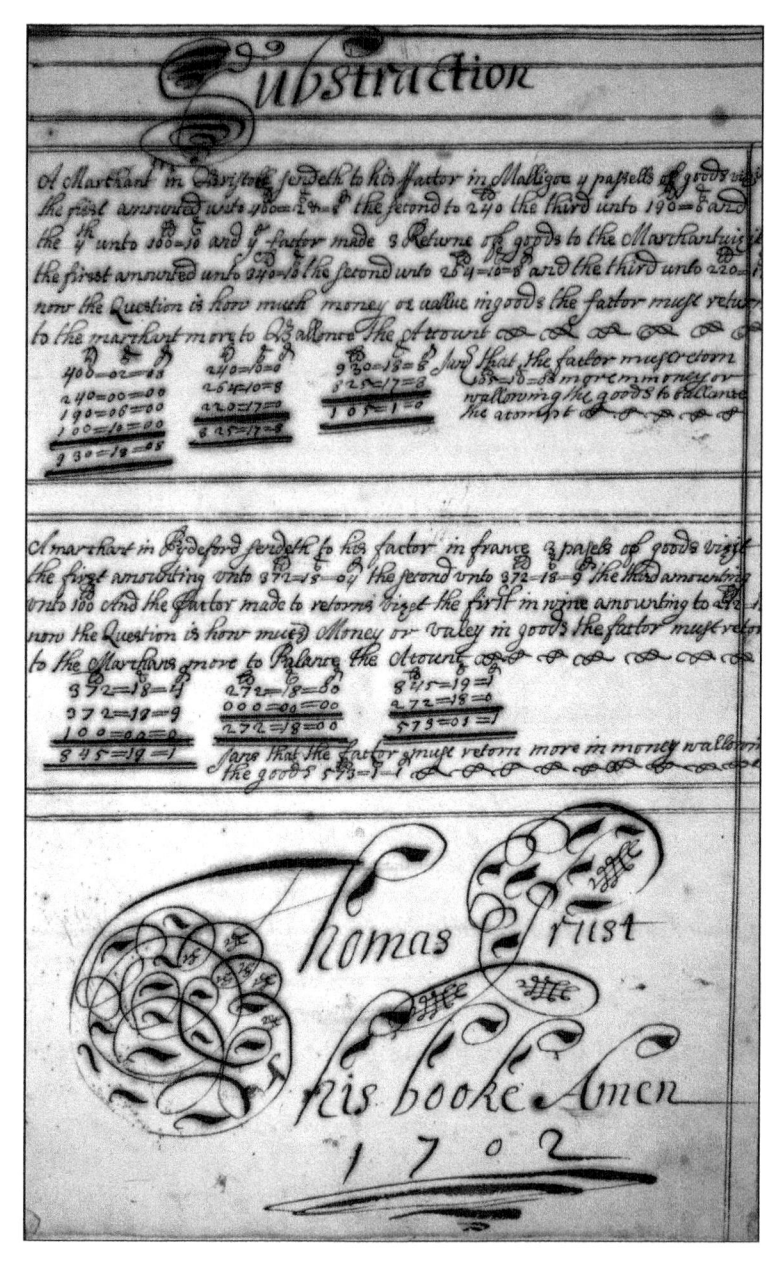

Figure 5.6. Statement and solution of two type problems (for subtraction), from Thomas Priest's ("Prust's") (1702) cyphering book (Appendix A, m/s #2).

William Webster, a London-based writing master and author of arithmetics at the beginning of the 18th century, pointed out that his first aim, in writing his popular 228-page *Arithmetick in Epitome: Or a Compendium of all its Rules, both Vulgar*

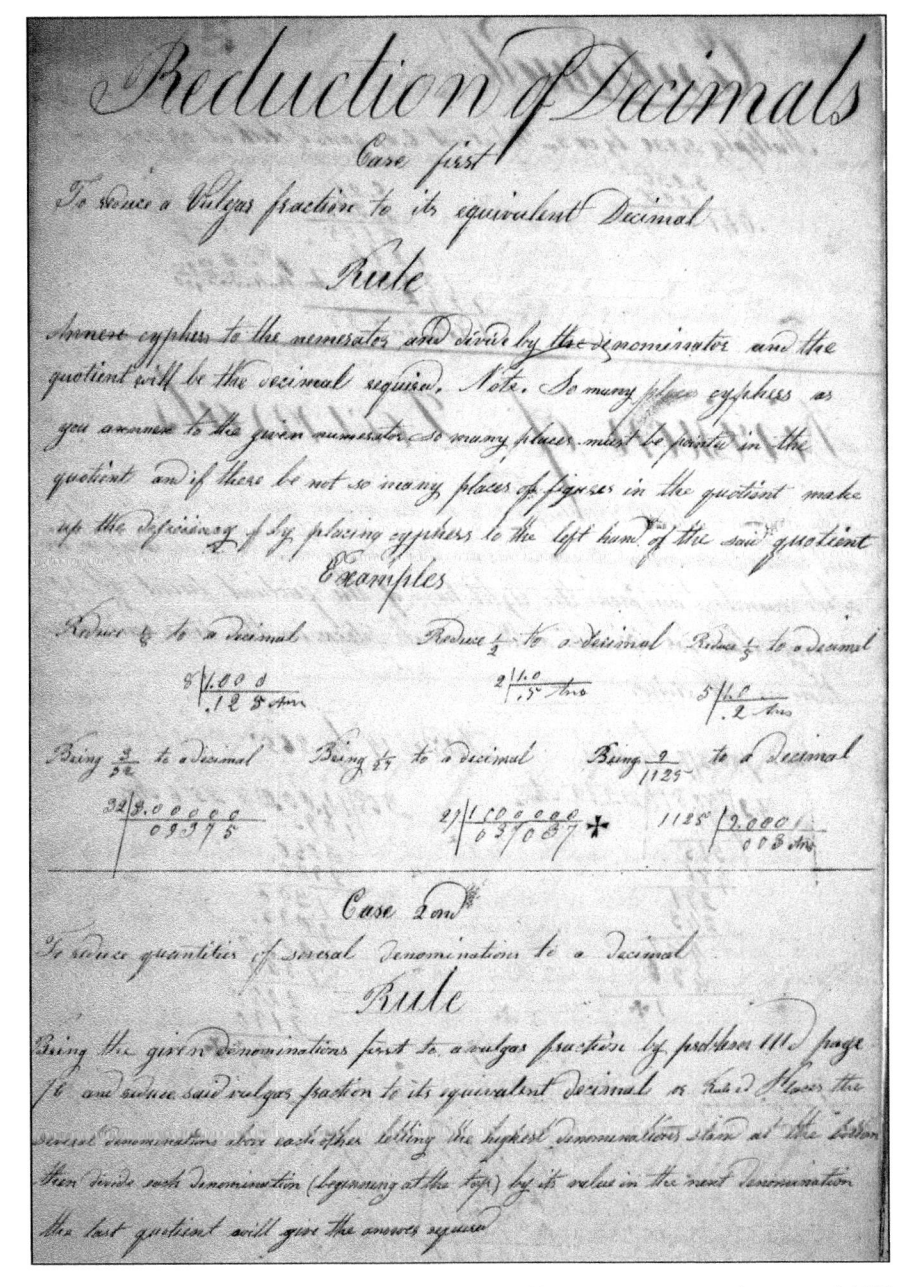

Figure 5.7. A page from the cyphering book prepared by an anonymous student around 1815 (Appendix A, m/s #79).

and Decimal (Webster, 1740), had been "to reduce to a pocket size what has sometimes been swelled to a folio" (p. vi). A similar genre was employed by other authors of arithmetics (see, e.g., Bonnycastle, 1788).

Authors of so-called "mental arithmetics" (often called "intellectual arithmetics"), however, usually adopted what they called an "inductive" genre (see, e.g., Colburn, 1821, 1822; Emerson, 1835; Smith, 1831), by which rules were not stated and students were expected to answer questions that were intended to help them identify rules and to see "reasons for rules."

Florence Yeldham (1936) commented on the genre adopted by John Mellis (1582) in the section that he added to the earlier (1543) arithmetic by Robert Recorde (in which Recorde had used the *abbaci's* more rhetorical dialogic genre). Mellis deliberately moved away from the genre adopted by Recorde. On a special title page that Mellis inserted before his lengthy new section toward the end of the 1582 edition of Recorde's original text, Mellis wrote: "The third part, or addition to this booke, entreateth of brief rules, called rules of practice, pleasant and commodious effects abridged into a briefer method than hitherto hath been published" (p. 345).

For better or for worse, this briefer genre characterized the style of writing adopted by most authors of printed arithmetics of the early 19th century, and we have observed it in almost all of the cyphering books that make up our PDS. This compact genre was introduced because business leaders preferred solutions to reckoning problems that arose in business contexts to be set out as briefly as possible (Walkingame, 1785). Although it is impossible to know the precise statistic, it seemed to us that about one-fourth of the solutions to exercises were copied without at least a minimal degree of understanding.

Many of the students who prepared the cyphering books regarded the contents as quite personal. To read the manuscripts is to be struck by the students' pride in their books. Of the 212 authors, 92 persistently wrote their names on pages and 22 wrote little poems from time to time. In diaries and biographies, authors' recollections of school days often made reference to cyphering: for example, Caroline Gilman (1852) wrote that her cyphering book had been "my pride and my mother's too—with what delight did she display those sums" (p. 317).

Seventy-one students wrote dates on the tops of pages on which they were writing. Of these 71 cyphering books, we could identify only 11 that were not prepared in their entirety during winter months. Most students who cyphered in the New England colonies/states, or in New York or Pennsylvania did so mainly in winter months, a conclusion which fits with how 19th-century U.S. schooling was portrayed by novelists (e.g., Eggleston, 1871) and with information recorded in personal diaries (e.g., Laughlin, 1845). However, students in southern states were much more likely to cypher throughout the year. Our analyses of cyphering books held at the University of North Carolina and at the College of William and Mary, left us in no doubt that the "Southern" cyphering books tended to be prepared throughout the year and tended to be longer than the "Northern" ones.

Relationships Between Mathematics Textbooks and Cyphering Books

Although some historians have claimed that students in 18th-century British settlements in North America learned arithmetic primarily from textbooks (see, e.g., Middlekauff, 1963; University of Michigan, 1967), this needs to be balanced against the likelihood that most students outside of the large cities did not own commercially-printed arithmetic books (Meriwether, 1907; Small, 1914). Some teachers might have owned their own commercially-printed arithmetic texts, but most of them based their methods of instruction on old cyphering books that they had copied or procured from other masters (Littlefield, 1904). Indeed, William Munsell (1882) claimed that well into the 19th century printed arithmetics were so scarce that many scholars never saw one. That appears also to have been true in relation to the use of printed arithmetics in schools in Great Britain in the early 19th century (Williamson, 1928).

When a parent or tutor in an early North American settlement decided that a student should progress beyond the simple counting and numeral recognition associated with "hornbook arithmetick," then that student usually did *not* purchase a personal copy of a commercially-printed arithmetic. Rather, he (for, more likely than not, the student would have been male) would procure sections of 8 or 16 pages of "rag" paper, and prepare a cyphering book. Topics would be announced with calligraphic headings. For each topic, introductions, rules, cases, type examples, and exercises would be entered, usually with as attractive penmanship as possible. Completed sections would, later, be sewn together to form "his book."

Alternatively, a student might have purchased a commercially-prepared book with blank pages—this was common in larger townships like Salem, in Massachusetts (Ellerton & Clements, 2009a). Some material may have been copied from printed textbooks on to slates (Monaghan, 2007; Monroe, 1917)—but this was not common in rural schools.

The Influence of Commercially-Printed Texts on the Preparation of Cyphering Books

American historians (e.g., Cajori, 1907; Cohen, 1982; Karpinski, 1925; Middlekauff, 1963) and British historians (e.g., Howson, 1982; Yeldham, 1936) have often pointed out that certain commercially-published arithmetics were used in the North American colonies in the 17th and early 18th centuries. These historians have drawn special attention to the use of arithmetics written by Robert Recorde (*The Grounde of Artes, Teaching the Worke and Practise, of Arithmeticke, Both in Whole Numbers and Fractions*, first published around 1543), Edmund Wingate (*Arithmetic Made Easy*, first published in 1630), James Hodder (*Hodder's Arithmetick, or, that Necessary Art Made Most Easie*, first published in 1661), Edward Cocker (both his *Arithmetick* and his *Decimal Arithmetick*, which appeared in 1678 and 1684, respectively), and William Leybourn (whose *Of Arithmetic,* was part of his *Cursus Mathematicus,* first published in London in 1690).

Recorde's *Arithmeticke* was used in at least one school in Massachusetts in 1653 (Karpinski, 1925, p. 174). Parts of Hodder's *Arithmetick* appeared in a text published in Boston in 1712, and the full text of the 25th edition of Hodder was published in Boston in 1719 (see Karpinski, 1980, pp. 38 and 41). Leybourn's (1690) *Cursus Mathematicus* is known to have been brought to Pennsylvania in the late 1690s by James Logan (Tolles, 1956).

Karen Michalowicz and Arthur Howard (2003) maintained, however, that "students of the 18th century rarely had a textbook," that those who studied arithmetic "wrote in a 'cipher' book," and that "the textbook was mainly for the teacher or for individuals who were self-taught" (p. 79). Walter Monroe (1917) quoted an educator in Boston around 1810 as stating that "printed arithmetics were not used in the Boston schools" until after he left there (p. 45). Rather, he said, teachers set "sums" for their pupils out of existing cyphering books.

Monroe (1912) claimed that throughout the 18th century many persons attempting to learn arithmetic never saw a commercially-printed arithmetic. Millard Kennedy and Alvin Harlow (1940), maintained that in local one-room schools in Kentucky and Indiana in the early 19th century, the invariably-untrained teacher had to deal with pupils aged from 5 to 25, a primitive fire-place, earthen floors, backless benches, no textbooks, little paper, a few slates, and no blackboard. Samuel Goodrich (1857) told a similar story about the rural school he attended in Connecticut in the late 1790s. In 1838, "Eliza B.," a Massachusetts teacher who received $1.25 per week for the seven-and-a-half-weeks that she taught in a one-room schoolhouse, filed the following report after she had left teaching at the school as a result of being unable to cope with its day-by-day demands:

> I am $15\frac{1}{2}$ years old. ... I wish to say the roof [of the schoolhouse] was all gone in one corner. You can see outside. The windows were all broken but we put paper over them. The floor was gone right under the bad roof. The fireplace does not heat except right in front of it. The wood was very wet at times as there is no woodshed. There are no conveniences for boys or girls. ... Children drink from a shared bucket of water, but the school is often so cold that the water freezes. ... The big boys took my bell so I could not call them in. Once inside the schoolhouse students made a "loud noyse" by scratching their slates. (Quoted in Zimmerman, 2009, p. 16)

According to Jonathan Zimmerman (2009), Eliza's school had "no blackboard, no teacher's desk, and not enough books to keep the school's 17 students occupied" (p. 16). In the 1830s, when Eliza taught, most schools were, in fact, one-room schoolhouses with one teacher (see, e.g., Wells, 1900). Even if a teacher knew and understood arithmetic—something not to be assumed—the teaching and learning of arithmetic under such conditions would have been arduous tasks.

Robert Clason (1968) stated that in early American schools, "pupils were provided with books of blank pages, called ciphering books, given a rule and a problem, and set to work" (p. 54). Walter S. Monroe (1917) asserted that textbooks were not generally used in the schools until well into the 19th century because teachers believed "in the efficacy of the method of teaching without a text" (p. 46). According to Monroe, teachers tended to believe that the cyphering approach to the teaching and learning of arithmetic was better than any other method. James P. Wickersham (1886 maintained that in early Pennsylvania schools, arithmetic "was done altogether without the aid of books," with "the sums being dictated by the master and worked out on paper" (p. 192). In summarizing what scholars have written on the matter, Susan Webb (2008) indicated that teachers in 18th-century North American schools "strongly stressed ciphering," that "arithmetic was usually taught without textbooks" and that "teachers had manuscript 'sum books' from which they gave out rules and problems in arithmetic" (p. 7).

After about 1840, the tide of opinion would turn against the cyphering approach. David Eugene Smith (1900) maintained that the approach was inefficient, and that good books saved "the time of pupil and teacher" (p. 139). According to Smith, as soon as textbooks became more readily available during the 19th century, there was no longer any excuse for a teacher to ask students to prepare cyphering books.

Smith seemed unable to conceive that a teacher could rationally believe that a cyphering approach might be preferable to a whole-class approach in which the teacher, armed with a commercially published textbook, took a more active, direct role. From Smith's (1900) perspective, the abandoning of the cyphering tradition in Europe and in America during the 19th century was a direct result of the large influence of Johann Heinrich Pestalozzi, who urged teachers to help young children grapple with "number rather than figures" (p. 85), through teacher-guided tasks involving real objects. Charles DeGarmo (1896) agreed with that assessment of the situation.

Peggy Kidwell, Amy Ackerberg-Hastings and David Roberts (2008), like Smith, believed that once cheaper textbooks became available, toward the end of the 18th century, it was inevitable that the cyphering approach would be abandoned. They wrote that before about 1750, books were relatively expensive, and only professors, tutors, or teachers owned textbooks used in classrooms. They went on to say:

> Students would write out specified sections in cipher or copybooks. Besides those men employed in education, only mechanics and others pursuing self-improvement purchased personal copies of manuals and textbooks for individual study. After about 1750, American college presidents required each student to purchase a textbook covering a range of mathematical subjects. Such compendia were commonly used in Great Britain. Thanks to a thriving Atlantic trade, the cost of such books had declined, making them affordable to students. (p. 5)

Kidwell et al. (2008) maintained that, at the school level, "pupils brought the arithmetics that their families had at home, so that in one room there might be as many different textbooks as children" (p. 6). This may have occasionally been true, but we have not found any documented evidence that it was *often* the case.

Kidwell et al.'s (2008) argument failed to take into account the power of the cyphering tradition on the expectations of parents, pupils, and teachers. Well into the 19th century the cyphering tradition remained the preferred method for most teachers of arithmetic because teachers regarded it as the most efficient way of assisting pupils to learn arithmetic. If, around 1800, a student brought an "arithmetic" from home to use at school, then in most cases that arithmetic would have been a handwritten cyphering book that had been, and was continuing to be, used from one generation to the next.

Kidwell et al. have underestimated the power of the belief among those inclined towards higher education in the 18th century that the process of using a quill to copy information from a known authority (such as an older cyphering book, or a well-regarded textbook) into a book that would become the writer's own had much educational merit. Thus, for example, we own an 80-page handwritten manuscript prepared by different members of the Terhune Family, of New Jersey, around the middle of the 18th century. The text in this manuscript was copied, in high calligraphic form, from John Love's *Geodaesia, or the Art of Surveying and Measuring of Land, Made Easie, . . .*, the first edition of which was published in London in 1688. Love's book was commonly available in and around New York at the time, but members of the Terhune family thought it worth the huge effort needed to copy it, by hand, with a quill. We have not included the Terhune manuscript in our PDS, for it does not fit our definition of a cyphering book, but the historical significance of its very existence should not be overlooked.

Analysis of the contents of the cyphering books in our PDS did not enable us to conclude that the cyphering approach produced high quality learning among students. Although some historians have commented on the power of the cyphering tradition to influence teachers and learners of mathematics (e.g., Michalowicz & Howard, 2003; Monaghan, 2007; Monroe, 1917; Smith, 1900; Wickersham, 1886), none has tackled the issue of why that tradition was maintained long after textbooks became readily available in the United States.

Gert Schubring (1987) has claimed that research in the 1970s established that "teaching practice is not so much determined by ministerial decrees and official syllabuses as by the textbooks used for teaching" (p. 41). Schubring added that although studying old school textbooks can be regarded as a quite traditional approach to the history of mathematics instruction, "we still seem to be far from reliable methods of analysis since the usual methodology of studying an isolated text or of formally comparing several books tends to neglect the social and cultural context of that utilisation as well as the national specificities of the respective school systems and the respective professional status of the teachers who used them" (p. 41).

Whether Schubring's (1987) contention is correct in relation to the history of mathematics education in Europe in the 18th and early 19th centuries is not a matter of concern in this book. However, we are confident that his theoretical position should not be applied to the school systems of the North American colonies, where hardly any mathematics textbooks were used in the schools. Even in the situation which prevailed in the United States between 1776 and 1840, when many teachers in schools with textbooks regarded those textbooks as being less important for assisting student learning than handwritten cyphering books, Schubring's theory would not seem to apply. In short, we believe that our analyses of data, and our interpretations of those analyses, in this present work, refute the thesis that the history of mathematics education in North America between 1607 and 1840 indicates that mathematics teaching and learning practices were determined by textbook authors and their textbooks.

In addition to our collection of cyphering books we also own a large collection of more than 1500 North American commercially-printed mathematics textbooks used in the eighteenth and 19th centuries. By comparing entries in cyphering books with those in printed texts we have concluded that at least 47 of the 212 cyphering books were at least partly based on material in published textbooks. Various editions of Nathan Daboll's arithmetics were used by at least 11 students, Michael Walsh's book was used by at least 6 students, and each of Zachariah Jess's and Nicolas Pike's arithmetics were used by at least four students. At least three students used Daniel Adams' texts, and another three used Charles Davies' books. Robert Gibson's *Surveying*, and John Bonnycastle's and Benjamin Greenleaf's arithmetics were each used by at least two students. Other authors whose books were used were Charles Brookes, John Gummere, John Love, Stephen Pike, Joseph Ray, William Slocomb, Thomas Smiley, John Stoddart, Elias Voster, and Jacob Willet.

Our conclusion, then, is that throughout the 17th and 18th centuries, and for part of the 19th century, the cyphering approach was likely to be used whenever and wherever North American students were learning "written arithmetic." We believe that exactly how published textbooks were used in conjunction with cyphering books varied from region to region, from context to context, from teacher to teacher, and from student to student. For example, Houser (1943) has shown that during the winters of 1824–1826 a young Abraham Lincoln prepared a cyphering book while attending a crudely-constructed "subscription" school in Southern Indiana. We examined that part of Lincoln's cyphering book held by the Houghton Library, at Harvard University. When preparing his cyphering book, Lincoln may have had access to a treatise on surveying by Robert Gibson, and to arithmetics, or pages from arithmetics, by Nicolas Pike or Nathan Daboll. Houser (1943) claimed that Lincoln's teacher may have deliberately removed pages from textbooks in order to enable different students simultaneously to copy different information from the same textbook. Whether that was indeed the case we cannot be sure.

We have identified the following five ways by which printed arithmetics may, or may not, have been combined with cyphering books:

1. *Neither the teacher nor the students owned a printed arithmetic, but the teacher owned a "master" cyphering book that was, perhaps, based on a printed arithmetic.* With this situation, the students could borrow the cyphering book from the master, or the master might dictate questions, or the master might write rules, cases, type problems, etc., directly into the students' cyphering books (Cubberley, 1962; Dyer, 1886; Karpinski, 1925; Lazenby, 1938; Meriwether, 1907; Monaghan, 2007; Monroe, 1917; Parker, 1912; Walkingame, 1785; Wickersham, 1886).

    William Fowle (1866) described how he studied arithmetic in a public school in Boston around 1810. Like Warren Burton (1833), Fowle recollected that students were not permitted to cypher at school until they were 10 or 11 years old. According to Fowle, printed arithmetics were not then used in Boston schools and, every day or so, the teacher wrote one or two problems in each student's cyphering book for the student to solve. Interestingly, Fowle recalled that writing and cyphering were never performed on the same day, and that students were permitted to copy solutions to problems from the cyphering books of more advanced students. Fowle maintained that while he was at school he never heard any teacher explain any principle of arithmetic.

2. *The teacher and students used the same single copy of an arithmetic textbook, and all students prepared their own cyphering books from that textbook.* This was a possibility when teachers who did not have many students lent their textbook(s) to their students (Houser, 1943; Jones & Coxford, 1970; Walkingame, 1785; Wickersham, 1886; Workman, 1789). Benjamin Workman (1789), writing during the early years of the new nation of the United States of America, stated that the purpose of his 224-page arithmetic was to present rules "so concise and simple that each may be written down or committed to memory in a few minutes, so that there is not a rule, illustration, or example throughout the book, but such as are absolutely necessary for the pupil to transcribe" (p. 11). The implication was that each student would copy a model problem, or introduction to a problem, and then pass the book on to another student. In that way, the teacher and all the students could use the same book.

According to Smith (1900), as late as the 1870s, the German educator E. Jänicke argued that because of the wide range of children's ability for arithmetic the teacher should use a printed book and refrain from whole-class instruction in the subject. Jänicke believed that teachers of arithmetic should move around their classes assisting individual children, and that under no circumstances should they dictate examples. Rather, each child should copy problems from the book and then silently and individually solve them. Jänicke also recommended that teachers themselves should master all the problems in any arithmetic text—handwritten or printed—that they used, in order that they would be as well prepared as possible to help their students.

3. *The teacher had a printed arithmetic textbook, and dictated problems from that textbook.* Robert Middlekauff (1963) claimed that "often teachers dictated problems which boys copied and solved" (p. 92). A variant on this occurred when the teacher actually wrote the questions to be solved on scrap paper, or on slates, or even directly into a student's cyphering book (Fowle, 1866; Monaghan, 2007; Munsell, 1882; Walkingame, 1785). Some teachers even ruled pages of students' cyphering books, and made calligraphic headings. It was this practice that was satirized by Charles Dickens, most notably in *David Copperfield,* in which frequent reference was made to how the tyrannical master, Mr Creakle, prepared the pages of his students' cyphering books.

4. *The teacher and students owned their own copies of the same arithmetic textbook, and this text became the basis for making entries in cyphering books.* Although this method was common, from the outset, in advanced colleges like Harvard and Yale, it was rarely applied in schools or evening classes before the middle of the 18th century. The method gradually became more common in day and evening schools after that (Munsell, 1882; Page, 1877; Walkingame, 1785; Wickersham, 1886).

5. *Working by themselves, learners copied rules and solved problems from printed textbooks.* Working by himself, Benjamin Franklin, who had not succeeded

in arithmetic when at school, subsequently, and famously, had no trouble in reading Edward Cocker's (1720) *Arithmetick* (Monaghan, 2007). Many authors of 18th- and 19th- century arithmetics (e.g., William Slocomb, 1831) claimed, in their prefaces, that their texts would benefit students—such as Samuel Laughlin (1845)—who were not in a position to have a teacher. Laughlin recalled that only two of about six of teachers in remote parts of Tennessee and Virginia between 1805 and 1815 had been able to help him learn arithmetic.

John Bonnycastle (1788), a British mathematician whose textbooks were used in many American colleges, wrote, in the preface to his arithmetic, that the "principal object of a work of this kind, should be to provide the learner with a proper set of rules and examples, so methodified and arranged, that they can be readily transcribed, and fixed in the memory, without any other assistance from the master than that of explaining the nature of the process, and examining the truth of the operations" (p. iii).

The educational thinking behind the idea was captured in William Pinnock's series of catechism booklets published in London during the first three decades of the 19th century. These booklets had titles like *A Catechism of Arithmetic, A Catechism of Geometry, A Catechism of Navigation, A Catechism of Land Surveying,* and *A Catechism of Algebra. A Catechism of Arithmetic* (1821), comprised 72 pages (with dimensions 5.5″ by 3.25″), and set out an *abbacus*-like sequence of rules, cases, type examples, and exercises, from numeration through to the rule of three direct.

IRCEE genre was evident in each of Pinnock's "catechisms." Many of the statements in the *Catechism* of *Arithmetic* were ready to be copied directly into a cyphering book and repeated in recitation sessions. On the last page, readers were referred to Pinnock's *Ciphering Book No. 2* for fuller accounts of some of the topics that had been summarized in the *Catechism of Arithmetic.* Although we have no evidence that Pinnock's series of catechisms were sold in the United States, it is interesting that, in 1835, two booklets titled *Cobb's Ciphering Book No. 1* and *Cobb's Ciphering Book No. 2,* were published in New York (Karpinski, 1980).

In Defense of the Cyphering Approach to Mathematics Education

It was easy for instructors at normal schools during and after the 1840s to condemn the cyphering approach, and it has also been easy for 20th- and 21st-century historians to draw attention to what they regard as the obvious weaknesses of the approach. Much has been written about children slavishly copying meaningless and largely irrelevant arithmetic that they did not understand, and about teachers harshly disciplining unwilling students who failed to memorize what they had dutifully copied into their cyphering books. While not wishing to elaborate too much on advantages of the cyphering approach, the following five points might be kept in mind by those critics who would roundly condemn it:

1. In its time, the approach was regarded as culturally appropriate. It had been passed on, from generation to generation, in many nations, as the best way to study arithmetic.
2. Printed textbooks were so expensive that many students could not afford to buy them. That comment did not apply to students in well-to-do families who sent their children to prestigious colleges (e.g., Gay, 1812).
3. High quality penmanship was regarded as a sign of a well-educated person. Thus, the writing-as-also-arithmeticke aspect of the cyphering tradition served an educative function that stretched beyond the realm of numeracy.
4. The expectation that whatever students wrote in their cyphering books should be thoroughly memorized and checked by teachers during one-to-one recitation sessions, had at least two educational advantages. The first was that it placed great pressure on students not only to memorize what they had written but also to have some idea of what it meant. The second advantage arose from the fact that much of what some students wrote in their cyphering books was, genuinely, important. At a time when there were relatively few books in the community, a personally-prepared cyphering book became an accessible repository for important information about numbers, measurement, and elementary business arithmetic. Students knew that they would be able to look in "their books" if, at some time in the future, they needed this information. This concept of a personal reference manual was particularly important for apprentices and others learning a trade. In their daily work many could not afford to purchase printed reference books, but they were able to refer to their own cyphering books.
5. Many students developed a genuine pride in, and ownership of, their cyphering books. And, since many of the books were passed on from sibling to sibling, and even from generation to generation, family pride was at stake. Children were expected to play their part in maintaining, and developing, a family tradition.

It is also relevant to point out, again, that even as late as the 1840s there still lingered, within high society, the feeling that the best scholars should normally take up classics. Many still believed that any time given by "top" students to developing mathematical competence was time lost from their working toward fulfilling the higher classical aims of education. We offer two examples, both from Phillips Academy in Andover, Massachusetts, to illustrate the deep-seated assumptions that resulted in the maintenance of this belief.

The first is in the form of a letter to parents and trustees by the Academy's Principal, Master Eliphalet Pearson, soon after Phillips Andover opened its doors for the first time in October 1780. Pearson wrote that at school assembly each morning, after devotions:

> A class consisting of four scholars repeats memoriter two pages in Greek Grammar, after which a class of 30 persons repeats a page and a half of Latin grammar; then follows the "accidence tribe," who repeat two, three, four, five and ten pages each. To this might be

added three who are studying arithmetic; one is in the rule of three, another in fellowship, and the third in practice. (Quoted in Adams, 1903, p. 49)

From the outset, the Phillips Academy School Board had decided that "preference be given to those scholars who are to be instructed in the learned languages" (quoted in Fuess, 1917, p. 75). Probably, the three students taking arithmetic (mentioned in the above passage) had been deemed to be unable to cope with the "more rigorous" curriculum in the classics. A bright 7-year-old, Josiah Quincy—who would one day become President of Harvard College—was forced to conform to the classics regime of the school. Quincy later recalled his "utter detestation" of the "abstract and the abstruse" in what he regarded as the over-emphasis on the classics during his early school days. He recalled that he often came home in tears, having been either "censured or punished" (quoted in Fuess, 1917, pp. 86–87).

This emphasis on the classics at Phillips Andover would continue. The "best" students routinely devoted most of their curriculum time to the classics, with "other" students being banished to the so-called "English Department" (Fuess, 1917). According to Claude Fuess (1917), "until 1866 the graduating class was composed only of classical pupils" and no-one ever proceeded from Phillips Andover to college who had taken the course provided by the English Department.

In 1847, James Eaton, who would become a well-known author of a widely-used arithmetic textbook series, was appointed Head of the English Department at Phillips Andover. In his first annual report, written in August 1848, Eaton was moved to complain:

> I have daily attended to from eight to ten recitations in the following branches: viz, geography, English, grammar with an analysis of poets, arithmetic, algebra geometry, trigonometry, mensuration, surveying, natural philosophy, astronomy, bookkeeping by double entry, reading and spelling. (Quoted in Fuess, 1917, p. 303)

At Phillips Andover, all weaker students were assigned to the English Department and studied a range of mathematical subjects through a cyphering approach. The best students were placed in the Classics division and studied little, if any, mathematics. The cyphering tradition operated within an educational climate that did not regard mathematics as important.

Ironically, much of the best teaching of mathematics took place in evening tuition classes frequented by students who had decided that they wanted to, or needed to, learn mathematics. Howson et al. (1981) referred to these schools as part of a "second tier" that had arisen because classical schools, in which Latin was the main language used, did not meet the needs of "mathematical practitioners or rechenmeisters" (p. 18).

The Nature and Role of Recitation

Much of early colonial mathematics education revolved around linking what was written in cyphering books or textbooks to the practice of recitation. Teachers would set individual students to work on copying certain rules, cases, model examples, or approved solutions to exercises, and would require the students to memorize what they had written—then, regularly (as often as twice a day), they would individually quiz students on what they were supposed to have memorized. This concept of recitation was applied to all areas of a curriculum—not just to cyphering. Charles Davies (1870), a best-selling 19th-century American author of arithmetics, summed up a viewpoint commonly held by teachers, students, and parents when he stated that "the recitation room was the final tribunal, and the intelligent teacher the final judge" (p. 8).

William Kinne (1831) was one of many arithmetic textbook authors who felt the need to offer a recitation guide for teachers. This comprised a list of questions that the teacher (or members of an examining committee) might ask during recitation related to arithmetic. Kinne (1831) also provided recommended answers. Thus, for example:

> *Recommended Question:* "What is arithmetic?" (p. v)
> *Corresponding statement in text:* "Arithmetic is the art and science of numbers." (p. 9)
>
> *Recommended Question:* "What are the four fundamental rules for its operation?" (p. v)
> *Corresponding statement in text:* [Arithmetic] "has for its operation four fundamental rules, viz. addition, subtraction, multiplication and division." (p. 9)
>
> *Recommended Question:* "To understand these, what is previously necessary?" (p. v)
> *Corresponding statement in text:* [To understand these] "it is necessary to have a perfect knowledge of our method of numeration or notation." (p. 9)

During the late 18th century and into the 19th century many authors (e.g., Adams, 1827; Smith, 1831; Sterry & Sterry, 1795) included prefaces, encouraging teachers to require students to link what was written in textbooks with what they wrote in their cyphering books and with recitation practices. They all condemned forms of recitation that relied solely on memorization, and most of them called for teachers to attempt to see if students understood what they had memorized. Some authors (e.g., Warren Colburn, 1822) called for teachers to ask students, during recitations, to explain how rules and methods could be applied in real-life situations.

At least two authors (Isaac Greenwood, in 1729, and Daniel Adams, in 1801) attempted to write textbooks which mimicked cyphering books. These authors stated the rules and cases, provided model examples, and then offered exercises with spaces left for students to write in the answers. The authors claimed that this would relieve teachers of much labor and would assist students. However, the approach did not work well, for two main reasons. The first was that leaving spaces for answers to be inserted meant that textbooks needed more pages, and therefore were more expensive than usual. The second reason carried a touch of irony: in fact, many teachers *wanted* their students to take a lot of time copying rules and model examples. In a class in which, say, 30 students aged from, say, 4 to 21 years, were mostly studying different subjects or topics, the teacher needed to buy as much time as possible to have any chance of coping.

Cohen's (2003) condemnation of the cyphering approach seems to have been based on her examination of some of the cyphering books held in the Houghton Library collection at Harvard University, and on Warren Burton's (1833) recollections of cyphering at a district school in the early 1800s. Burton stated that when he began to cypher, at the age of 12, his family purchased a new copy of Daniel Adams's *Arithmetic* to assist him with his new endeavour. Although between 1801 and 1827 all editions of Adams's *Arithmetic* left spaces for students to answer questions in the book, the young Burton copied rules from the textbook into his cyphering book.

Writing 20 years after the event, Burton made clear that being given the opportunity to create his own cyphering book represented an important *rite de passage* for him. Both he and his family believed that this would take him to the next stage of his education. Indeed, Burton perceived the creation of his own cyphering book to have been a far more important event in his life than being given, and being able to use, a new copy of Adams's *Arithmetic*.

According to Keith Hoskin (1994), during the first half of the 19th century, daily recitations at West Point Military Academy were based on nominated sections from prescribed textbooks. However, the sections from the textbooks were entered into cyphering books by the students, and it was these cyphering books that the students retained, and sometimes used, after they had departed college.

Chapter 6
Ending a 600-Year Tradition: The Demise of the Cyphering Approach

Abstract The question of why the influence of the cyphering tradition on school mathematics in North America fell away, so quickly, between 1840 and 1861 is considered. Three reasons are given: first, the introduction, in the 1840s, of externally-set written examinations in mathematics meant that no longer were cyphering books regarded as the most important evidence in relation to teaching and learning efficiency; second, normal schools spread rapidly after 1840, and those who were trained in the normal schools were taught that teaching and learning methods based on the cyphering tradition were hopelessly inadequate; and, third, education administrators, such as Horace Mann, placed pressure on schools to adopt whole-class methods of teaching mathematics, in which the teacher played a much more active role than ever before.

Warren Colburn and the Challenge to the Cyphering Tradition

In the 1820s serious questions began to be asked about the purposes of school arithmetic, and there seemed to be a distinct possibility that teachers and schools would move away from the 600-year-old cyphering-book, *abbaco* tradition. Some historians have suggested that the ferment of ideas began in 1821, when Warren Colburn, a recent Harvard graduate, published a little book aimed at assisting teachers to help young children (from around six years of age) to learn the first principles of numeration, counting, and the four operations. Through the title of his book (*An Arithmetic on the Plan of Pestalozzi, with some Improvements*), Colburn (1821) acknowledged the influence of Johann Heinrich Pestalozzi, the great Swiss/German educator, on his thinking.

Although Colburn only taught in schools for a few years, *An Arithmetic on the Plan of Pestalozzi* was destined to become one of the most influential texts in the history of U.S. education (Cajori, 1890; Doar, 2006). According to George H. Martin (1897):

> Colburn's book came into the schools as refreshing as a northwest wind, and as stimulating. It was eagerly seized upon by the more intelligent teachers. Its use was a mark of an intelligent teacher, a sign of life from the dead. Embodying the principles of the new education, it wrought a revolution in the teaching of arithmetic, and it determined the character of all subsequent text-books. (p. 40)

That statement summarizes the standard interpretation of how the cyphering-book tradition lost its impact in the United States. This standard interpretation (as also stated, for example, by Breed, Overman and Woody (1936) and Monroe (1912)) points to how American teachers, inspired by Pestalozzi, began, through Colburn's writings, to use inductive, mental, methods that invited students to construct their own mathematical knowledge. This not only resulted, so the tradition goes, in students beginning to learn arithmetic at about six years of age (rather than at 10 or 11), but also in more students understanding more arithmetic than ever before. Furthermore, females began to study arithmetic in much greater numbers than at any previous time.

According to the standard interpretation, a new breed of Colburn-inspired teachers quickly energized the teaching and learning of arithmetic. Among other things, these teachers adopted more inductive approaches so that, right across the nation, teachers began to ask young learners to figure out answers mentally to carefully sequenced sets of questions, and then to articulate those answers in well-formed sentences. In this way, the teacher played a far more important role in facilitating arithmetic learning than ever before.

We believe that available data do not support this standard interpretation. The first relevant indicator is the distribution into decades of cyphering books in the PDS. In Figure 5.1, it can be seen that for the decades 1820–1829 and 1830–1839 the numbers of cyphering books are greater than the numbers for the decades 1800–1809 and 1810–1819. Although it is recognized that Figure 5.1 is not based on a sample of cyphering books selected randomly from across the nation and from across periods of time, and that, in particular, the number of school-age children in the United States was steadily increasing between 1800 and 1839 (see Figure 5.2), the data summarized in Figure 5.1 are not consistent with the standard interpretation. It appears to have been the case that, following the publication of Colburn's (1821, 1822) early books, there was no sudden and dramatic decrease in the use of cyphering books in the 1820s and 1830s.

It is not our intention here to question the importance of Colburn's (1821, 1822) texts or of other arithmetic texts written in the 1820s and 1830s that were directed at teachers of young children (e.g., Goodrich, 1833). Rather, it seems that most U.S. teachers who taught arithmetic to students aged ten years or more continued to use the cyphering approach throughout the period 1820–1840. Although Colburn may well have influenced generations of teachers who taught young children up to ten years of age, there is little evidence that he had a strong influence on teachers of older students. He did write an arithmetic textbook, and an algebra textbook, for older students (Karpinski, 1980), but these later books were not nearly as popular, or as influential, as the arithmetics he directed at teachers of elementary school children.

There are first-hand testimonies to the distinctiveness of Colburn's approach to arithmetic education, but often those making such testimonies point out that Colburn's approach was opposed by many teachers. For example, James Freeman Clarke, who used one of Colburn's arithmetics at the Boston Latin School in the 1820s, commented, many years later, that "this admirable book was soon banished from the schools by the pedants, who thought that whatever was interesting must be bad" (Clarke & Hale, 1892, p. 278). As far as we can ascertain, none of the 212 cyphering books in the PDS was based on any of Colburn's books.

Monitorial Systems for School Mathematics

Colburn was not the only person, during the period 1820–1840, to try new ideas for teaching young children arithmetic. In Boston, for example, William Bentley Fowle (1795–1865) was responsible for the introduction of a monitorial system of instruction into the public and private schools of Boston. In 1843, Fowle, a friend of Horace Mann, became a member of the Massachusetts legislature. He showed considerable political flair in his advocacy of teaching approaches in which older pupils taught younger students according to a carefully developed scheme (Paolitto, 1976). For example, in 1846, Fowle (1866) told a gathering of teachers:

> I am confident from my own observation that nearly all occasion for severe discipline in schools is owing to the fact that most children in school really have nothing to do for a very large part of the time. In a school of 50 scholars, no-one is entitled to more than three minutes and a half of the teacher's time in a day. The child must sit still, if he can, nearly three long hours, and the teacher is held to be no teacher, and his school no school, if children so situated—play. (p. 16)

Statements such as the above appealed to many teachers, and caused them to think that the cyphering/recitation approach was inefficient. Fowle (1866) made his beliefs on the matter public when he wrote a lengthy chapter criticizing traditional approaches to teaching arithmetic.

During the period 1815–1840, various forms of the monitorial system were tried and found wanting in different states (Cubberley, 1962). Sometimes those who supported monitorial schemes believed that cyphering should be a part of the curriculum at monitorial schools (e.g., Andrews, 1830), but the overall effect of the move towards monitorial approaches was to undermine public confidence in the cyphering tradition. Joseph Lancaster (1805) claimed that his monitorial system, by which advanced or older pupil monitors taught small groups of less advanced pupils, wholly superseded the cyphering approach (Fox, 1809).

The Continuing Emphasis on Cyphering 1820–1840

But the cyphering approach continued to be widely used during the period 1820–1840. Confirmation of the continuing emphasis on the approach can be found in the form of documents, deriving from the mid-1820s, which suggest that there were many influential persons who clung to the belief that the cyphering approach to teaching arithmetic and writing was both effective and efficient. Dorothea Dix (1826), a teacher, asserted that cyphering was "of great importance to everyone, but especially so to children, for by it they not only acquire what they would be obliged to learn in more advanced age, but their understandings are improved, and their memories strengthened, while at the same time they are taught to study, and fix their minds upon one object" (p. 76).

Between 1825 and 1830, Dudley Leavitt, a self-trained scientist and a well-known author and publisher of almanacs, who conducted a private school in Meredith, New Hampshire, wrote a series of arithmetics aimed at helping teachers and students to cypher (Leavitt, 1826a, 1826b, 1830). His approach to cyphering was made particularly clear in the following introductory statement in one of his textbooks, which carried the title *The Teacher's Assistant, and Scholar's Mathematical Directory, Containing a Solution of the Most Difficult Questions in Pike, Walsh, Adams, Daboll and Robinson's Arithmetics, and Flint's Surveying* (Leavitt, 1830):

> Respecting the solution of difficult questions in this book, the first part of most of them is all that is set down, and is included in a quotation: that being all which is necessary as a directory for finding the question at large in any author referred to.
>
> Example: Suppose you are ciphering in Barter in Walsh's *Arithmetic*, and come to this question—"C and D barter; C of 7 s makes 6 s.8d, D of 7 s.6d makes 7 s.3d: who has lost most, and by how much per cent?"
>
> Turn to the *Solution of Arithmetical Questions in Walsh's Arithmetic*, given in this Directory, and under Barter, run your eye over the beginning of the question and you will soon find the one corresponding, and the method of solving it: and so of any other. (p. 7)

Leavitt stated that his 1830 text, was "composed and collated for the use of schools" (p. iii).

Leavitt's (1826a) abridgment of Nicolas' Pike's early arithmetic was set out according to the IRCEE genre and was probably written with the needs of students preparing cyphering books in mind. The same was true of Leavitt's (1826b) *A New Ciphering Book, Adapted to Pike's Arithmetick, Abridged. Containing Illustrative Notes, a Variety of Useful Tables*.

For Leavitt, the increasing availability of arithmetics did not raise the question whether students should continue to prepare cyphering books. From his point of view, the preparation of cyphering books was an essential ingredient in high quality mathematical education. He wanted to make it easier for teachers to assist students to prepare more helpful cyphering books by giving them more ready access to authoritative arithmetical information and methods.

Be that as it may, Leavitt's decision to interpret cyphering as a process whereby students would merely copy written solutions to problems set in some of the more popular arithmetics cheapened the concept of cyphering. Although, in the past many students had copied solutions into their cyphering books from older cyphering books, owned by others, at its best the cyphering tradition had always been something more than copying. For centuries it had been assumed that students would solve the exercises by themselves, and would then have their tentative solutions checked by masters during the recitation process. Then, and only then, would they be permitted to enter their solutions neatly into their cyphering books. Leavitt was a highly regarded teacher of mathematics, but his approaches lessened the educational value of cyphering and could have provided ammunition for educators, like Horace

Mann, who wanted to see the cyphering approach dismissed from schools in the United States.

Lyman Cobb (1834) criticized writers like Bonnycastle, Daboll and Pike for giving exercises and rules without explanation. He also criticized other writers, like Colburn and Emerson, for providing neither rules nor explanations, "so that the scholar must perform all his operations mentally, not having any rule to direct him, and consequently can have no theoretical knowledge of the subject by which to be able to work questions in after life, depending entirely on the teacher for practical information only" (p. 23).

Cobb (1835) stated, in his preface to *Cobb's Cyphering Book, No. 1,* that he had illustrated how answers should be written—the actual penmanship was shown—for the following six reasons:

1. Scholars may need to learn to write figures. Very many scholars, and even adults, write words quite well, but figures badly.
2. They may need to learn to set out their sums accurately.
3. They may need to be induced to keep their books nicer and cleaner, and take more pains to preserve them, than they would common writing books used for setting sums.
4. That scholars are frequently disheartened and consider their time and labor lost, if, after they have toiled an hour or two, perhaps longer, in working an intricate sum, they, with one brush of the hand or sponge, strike their work from the slate without writing it in a book for preservation; and, again, the writing of the sums impresses them more deeply upon the mind than the mere working of them possibly can.
5. These cyphering books will answer as a reference in after life, when cares and business shall have effaced or obliterated from the minds of those who have used them, the particular working of these sums.
6. That these books will answer a good purpose for reviewing the scholars, by the teacher's requiring them to state the reasons for the manner in which each sum has been worked. (p. 2)

Cobb added that "with these rules, the first sum or sums under each rule have been worked out and printed in *script* type; and every other sum or question has been printed without rule, explanation, or answer, with such a blank after each sum, or questions, that the working of the sum and the answer can be inscribed or copied at full length from the slate" (p. 2). Notice that, unlike Leavitt, Cobb assumed that the students would have expended intellectual effort in preparing their solutions on their slates.

In 1835 a pre-prepared cyphering book with 172 blank, unlined pages, each with dimensions 12.5″ by 7.5″, was published in Pennsylvania by a certain Reuben Chambers. The interior pages were of high quality paper, and Chambers' idea was to save students the bother of having to construct and sew together their own cyphering books. The most telling part of Chambers' (1835) pre-prepared cyphering book was the following "note to teachers," printed on the back cover, regarding the use of cyphering books:

> It is common with young beginners, at cyphering, to neglect or be ignorant of the proper use of the capital letters. Which (sic.) ought to be used to begin every sentence, and proper names of persons or places, and also of the use of the different points and stops before them, even sometimes when they have the advantage of a correct printed copy before them.

> We often see in the finished cyphering books of young arithmeticians great irregularity, carelessness, and want of judgment in setting down their sums; often by not setting the figures in a proper situation under each other, and often by drawing the black horizontal lines, between the rows or lines of figures, much longer than necessary, and often in an oblike (sic.) direction, instead of square or at right angles to the page. The correct method of beginning every new paragraph as far from the margin, as one line is from another, ought also to be attended to, as the following paragraph.
>
> All abbreviations of words, whether in arithmetic, or any other place, should be marked with periods at them, as Dols. for dollars, Cts. for cents, Cwt. for hundred weight, Qrs. for quarters, Lbs. for pounds, &c, &c. These things ought all to be strictly attended to by the scholar, with as much care and exactness as is bestowed in the figure work of the questions. Blots and mistakes ought also to be avoided as much as possible; that neatness and a uniform improvement may appear throughout the book.
>
> The teacher should loose (sic.) no opportunity for acquainting his pupils with the usefulness and necessity of attending to the above observations; for if they should not have the opportunity of studying grammer (sic.), this will assist them much in their business transactions in after life. (Chambers, 1835, back cover)

Chambers assumed that the cyphering approach would continue to be widely used in the schools. However, he did not place much educational value in the calculational and reasoning aspects of arithmetic. His statements suggested that, even in the mid-1830s, some still believed that writing and arithmetic education should continue to be intimately linked.

Warren Burton (1833) recalled that when he was a student in a Boston school around 1812 he used slates as an aid to cyphering. A student could work out a draft solution to an exercise on a slate, get this checked for accuracy by the teacher, and then write the checked solution neatly into his or her cyphering book. No paper was wasted in the process. In 1825, the Reverend Adam Marshall, a Jesuit priest employed to teach arithmetic to the midshipmen of the United States ship, the *North Carolina,* during a mission to the Mediterranean, deposited in his room on the ship, "30 slates and 100 pencils" (Durkin, 1942, p. 155).

Although blackboards had been used in some American schools early in the 18th century (Monaghan, 2007), it was not until the 1830s that they were commonly used. The February 1839 issue of the *Connecticut Common School Journal,* for example, argued that all teachers should use blackboards and that pupils should also use them "for demonstrations and illustrations" (p. 5). Although one might have expected this to facilitate the practice of cyphering, teachers trained in normal schools would be shown how to use the blackboard for whole-class teaching of arithmetic that deliberately avoided the use of cyphering books (Holbrook, 1859).

The Demise of the Cyphering Approach

One should not overemphasize the amount of cyphering that took place in early U.S. schools. Most students who prepared cyphering books were at least ten years old, and probably less than 20 percent of all school children ever prepared a cyphering book. Most of the female students who stayed on at school after gaining an elementary education usually spent much of their time on sewing, or on other tasks which were regarded as more "feminine" than cyphering. For students who did proceed to cyphering, the amount of cyphering that they actually did was often quite limited. Many of the boys, in particular, attended schools for no more than 16 weeks each year—from early December through late March. Between April and November they worked on family-related tasks. However, some older students, after leaving school, decided to learn cyphering through private tutors and special classes for apprentices.

In 70 of the 212 cyphering books (i.e., 33 percent) from the PDS, students actually wrote the dates on which they were cyphering on individual pages and in 55 of these 70 cases the cyphering took place during the December-through-March period. In rural areas this pattern of disjointed school attendance continued until almost the end of the 19th century. This conclusion is based on dates indicated on pages of many manuscripts within the PDS, and is supported by educational statistics when these are available. Thus, for example, James P. Wickersham (1886) pointed out that in the early 1830s there were at least 400,000 children in Pennsylvania aged between 5 and 15, but only about one-third of them attended school in any particular year, and hardly any boys attended during the months from April through November.

Gradually, during the 19th century a free elementary education became available for all children, in all states, and arithmetic came to be given more curriculum time than any other subject (Smith, 1911). In short, the idea of "arithmetic for all," within a free public school system became a reality. But, in the early 1840s many adults did not believe that all children needed to study arithmetic beyond the counting they learned at dame schools or elementary grade schools.

After the 1840s the concept of grade schools, in which students were grouped in rooms according to age, quickly became common in well-populated urban areas (Cajori, 1890; Tyack & Tobin, 1994). This facilitated whole-class teaching to grades of children who, supposedly, were at about the "same level." In the less populated rural areas, however, children of all ages were still educated in one-room schoolhouses, where whole-class teaching was much less feasible. The tradition of the "one-room school-house" would remain alive in these rural schools for another century. Relatively few of the students who attended these one-room schools would proceed to higher academic studies—Myrna Grove (2000) estimated that as late as 1900, "only five percent of one-room school graduates proceeded to urban high schools" (p. 75).

Why Did the Cyphering-Book Tradition Lose Favor During the Period 1840–1865?

Our analyses of our PDS has suggested that as late as 1840 the cyphering tradition still largely influenced how arithmetic and other forms of arithmetic were taught and learned in the United States of America. By 1865, however, most teachers of mathematics taught mathematics in ways that were not consistent with the cyphering tradition, and indeed not only do we not have a cyphering book in our PDS that was prepared after 1861, but we have not found, in any of the collections of cyphering books that we have examined, a cyphering book prepared in the United States after 1861. The question arises why and how this change occurred so quickly, given that the tradition was centuries old?

During the period 1840–1860 a range of new factors influenced what was included in US mathematics curriculum documents and textbooks—the intended curriculum—and how it was studied—the implemented curriculum. It will be useful to discuss some of those influences, and assess their impact.

Greater availability of reasonably-priced textbooks. The 512-page first edition of Nicolas Pike's *The New Complete System of Arithmetic, Composed for the Use of the Citizens of the United States*—the first major arithmetic textbook authored by a citizen of the United States—was published in 1788. It was large, expensive, and more suited to college students and to those preparing for college than to those simply trying to learn the basics of arithmetic for living (Ellerton & Clements, 2008). Nevertheless, Pike's textbook, like the English-language textbooks by Noah Webster (Ellerton & Clements, 2008), heralded the beginning of a school textbook industry in the United States (Small, 1914).

Between 1788 and 1840 many relatively inexpensive arithmetics written by US citizens would be published. Some of these would go through many editions, and would be more suited than Pike's arithmetics to the needs and abilities of school children. The most popular were written by Zachariah Jess (first edition 1799, and 12 editions by 1819), Nathan Daboll (first edition 1800, and 51 editions by 1823), Michael Walsh (first edition 1801, and 19 editions by 1832), Daniel Adams (first edition 1801, and 10 editions by 1818), and Stephen Pike (first edition 1811, and 6 editions by 1829). From the 1820s, authors such as Warren Colburn, Frederick Emerson, Roswell Smith, Charles Davies and Joseph Ray began to write series of mathematics texts—mainly arithmetics—with the idea that students would progress from one text to another, or that different students of the same age might be better suited to different texts (Karpinski, 1980).

The extent to which the publication of these new textbooks contributed to the demise of the cyphering-book tradition is difficult to determine. At first, authors tried to write their books so that they complemented, rather than competed with, the cyphering tradition (Dyer, 1886). On the title page of Francis Walkingame's (1785) *The Tutor's Assistant: Being a Compendium of Arithmetic and a Complete Question-Book*, the author was described as a "writing master and accomptant" (p. i), and Walkingame devoted considerable space in his preface to explaining how his book should be used to complement the cyphering approach. Later editions of Walkingame's text were still widely used in Canada during the first half of the 19th century. John Bonnycastle (1788) stated in the preface to his *Arithmetic* (which was not only widely used in England but also in some North American colleges) that "the principal object of a work of this kind, should be to provide the learner with a proper set of rules and examples, so methodified and arranged, that they can be readily transcribed, and fixed in the memory" (p. iii).

Daniel Adams (1801) wrote, in his preface, that his book would relieve masters "of a heavy burden of writing out rules and questions, under which they have so long labored, to the manifest neglect of other parts of their schools" (p. iii). Later in his preface, Adams stated that the blank space after each example in his arithmetic text "was designed for the operation by the scholar, which being wrought upon a slate, or waste paper, he may afterwards transcribe into his book" (p. v). Clearly, Adams was fully aware of the strength of the cyphering-book tradition (Figure 6.1 shows a reproduction of page 32 from Adams' (1822) book in which "blanks" were completed by the owner.). Adams (1827) later wrote a new version of his book in which he changed the format so that students were no longer given space to write in answers.

During the period 1840–1865 there was a decline in teachers' expectations that cyphering books be created by students, and a sharp increase in the number of commercially-printed school arithmetics and other mathematics texts (Karpinski, 1980). By 1870, more and more teachers had decided that each student learning arithmetic needed a textbook, and teachers rarely expected students to generate cyphering books of the old kind (Meriwether, 1907). Students now purchased commercially-printed, lined exercise books, and teachers expected them to solve problems, set in the textbooks, in these exercise books. Excellent penmanship and calligraphy were not as important as they had been in earlier times. We have not found any evidence to show that from the 1870s the cyphering approach was used in any North American schools.

Cyphering books lost their relevance for student and teacher evaluation. As noted earlier, throughout the 18th and the early 19th century the cyphering book played a vital role in the assessment of the quality of both the students' learning and the teachers' instruction. At the end of each semester a public examination of a school's work was held, and on these occasions cyphering books were inspected by a school committee and by parents. A teacher's future career may have depended on the attractiveness of the cyphering books, and hence the emphasis on excellent penmanship and calligraphy was not surprising.

From the 1840s onwards, individual assessment by public committees was gradually replaced by assessment based on written examinations. This trend had begun in England and on the Continent in the late 1700s, but was not accepted by key education authorities in the United Kingdom until the 1830s, and in the United States until around 1840 (Henry, 1843; Roach, 1971; Rotherham, 1852; Watson & Kandel, 1911). The introduction of external written examinations would prove to be the death knell for the use of cyphering books in U.S. schools, and how and why that proved to be the case is worthy of some attention here.

As Secretary to the Massachusetts Board of Education, Horace Mann criticized the committee system that was used for evaluating the work of students and teachers (State of Massachusetts, 1848). Mann (1845/1925) claimed it would be better to examine students through externally-set, written examinations, because for any subject, all students taking a written examination in a particular grade, at different

32   SIMPLE DIVISION.   SECT. I. 4.

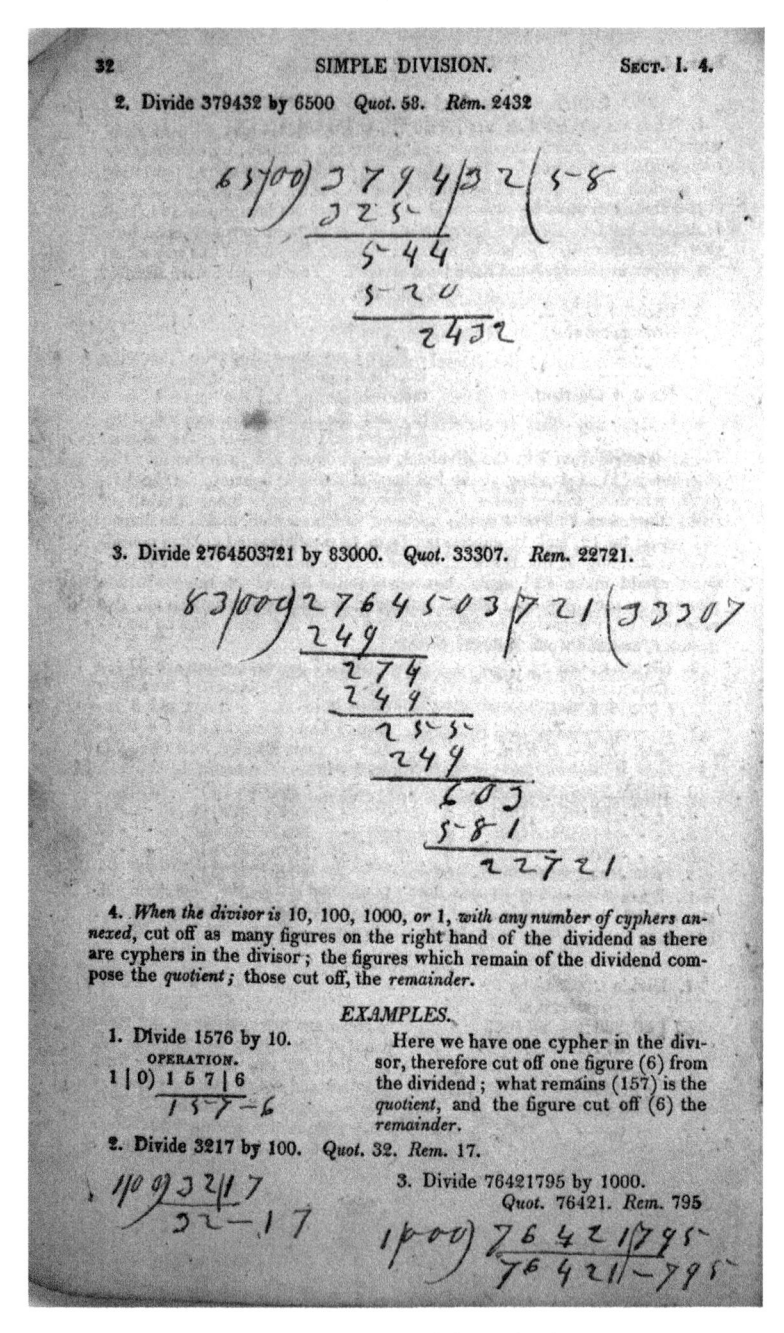

2. Divide 379432 by 6500.   *Quot.* 58.   *Rem.* 2432.

3. Divide 2764503721 by 83000.   *Quot.* 33307.   *Rem.* 22721.

4. *When the divisor is* 10, 100, 1000, *or* 1, *with any number of cyphers annexed*, cut off as many figures on the right hand of the dividend as there are cyphers in the divisor; the figures which remain of the dividend compose the *quotient*; those cut off, the *remainder*.

EXAMPLES.

1. Divide 1576 by 10.
   OPERATION.
   1 | 0) 1 5 7 | 6

   Here we have one cypher in the divisor, therefore cut off one figure (6) from the dividend; what remains (157) is the *quotient*, and the figure cut off (6) the *remainder*.

2. Divide 3217 by 100.   *Quot.* 32.   *Rem.* 17.

3. Divide 76421795 by 1000.   *Quot.* 76421.   *Rem.* 795.

Figure 6.1. Page 32 from Adams (1822), with "blanks" completed by an owner.

schools, could answer the same set of questions. A common measuring stick would be created for assessing the learning of individual students, thereby making assessment of the quality of student learning and of the effectiveness of teachers more accurate. According to Mann, if the written tests were set externally and the process administered by disinterested experts, it would become difficult for teachers to interfere in the examination process in educationally significant ways.

Mann's arguments for written examinations were presented in an explosive political climate in which his work as Secretary to the Massachusetts Board of Education was under attack by a group of "Boston schoolmasters" who were defending traditional curricula, traditional classroom organization, and traditional committee assessment (see Boston schoolmasters, 1844; Caldwell & Courtis, 1925; Katz, 1968; Kilpatrick, 1992; State of Massachusetts, 1848). In 1845, Mann arranged for members of his Board of Education to test Boston schools using a battery of written tests prepared by the Board. These tests had not been seen by the schoolmasters before the examinations were conducted. The examinations were administered by Board members, who rushed from one school to another to administer the same examinations.

The written arithmetic questions, and performances of the students on individual questions, are shown in Table 6.1, which provides rare data on the arithmetic performances of students trained under the cyphering approach. There were 8343 students enrolled in the 19 schools in which the examinations were administered, but only 308 students (average age, 14 years and two months) took the tests (Caldwell & Courtis, 1925). Mann referred to the 308 students who took the test as the "brag" students of the schools (Caldwell & Courtis, 1925, p. 52), and indicated that the examinations had been designed for students of this category. All 308 students had learned their arithmetic in the publicly-supported "writing" schools of Boston (Caldwell & Courtis, 1925) in which the cyphering approach was employed.

After the students' responses were graded, Mann claimed that the students' arithmetic performances were poor. The schoolmasters quickly responded by claiming that the Board's examinations were woefully invalid, in the sense that questions did not bear a strong relationship with the arithmetic that they had asked their students to learn.

Otis Caldwell and Stuart Courtis (1925) argued, correctly from our own perspective, that the schoolmasters were correct when they asserted that the arithmetic examination was a very poor test. There were four examples involving computation and six problems that required students to identify and sequence a set of operations. Only one of these problems was within the children's capabilities. For two problems, none of the 308 children obtained a correct answer; for three other problems, the numbers of correct answers were 1, 2, and 3 (out of 308) respectively. Such results suggest that the examiners were not familiar with the work of the schools and that they greatly overestimated the attainments of the children. Mann and his committee believed, however, that since only "brag students" took the test, all of the questions were fair.

All 308 students had learned their arithmetic through a cyphering approach (Boston Schoolmasters, 1844), complemented by Frederick Emerson's (1835)

Table 6.1

Questions on the First U.S. Externally-Set Arithmetic Paper Administered to 308 Public School Students, with Numbers and Percentages Giving Correct Answers (Caldwell & Courtis, 1925)

Question	Number of Correct Responses ($n = 308$)	% Correct
1. How much is 1/3 of nine hours and 18 minutes?	288	93.5
2. What part of 100 acres is 63 acres, 2 roods and 7 rods?	282	91.6
3. What is the quotient of one ten thousandth divided by ten thousand? Express the answer in decimal and vulgar fractions.	170	55.2
4. A stationer sold quills at 10s 6d per thousand, by which he cleared 1/3 of the price—but, the quills growing scarce, he raised the price to 12s per thousand. What percent would he clear by the latter price?	9	2.9
5. Suppose A owes me $100 due at the end of three months and $100 due at the end of nine months, and he agrees to give me a note for $200 payment at such a time that its present worth shall be the same as the sum of the present value of the two first-mentioned notes. How long after the date must the note be made payable?	0	0
6. A man has a square piece of ground which contains one quarter of one acre and a quarter on which are trees which will make wood enough for a pile around on the inside of the bounds of the land 3 feet high and 4 feet wide. How many cords of wood are there?	0	0
7. A sold goods for $1500 to be paid for one half in six months and one half in nine months. What is the present worth of the goods, interest being at 7 percent?	147	47.7
8. A merchant in New York where interest is 7 percent gives his note dated at Boston, where interest is 6 percent for $5000 payable at the Merchants Bank, Boston, on demand. Thirty days after the date of the note, demand is made. A year after demand, $200 is paid on the note. What sum remains at the end of two years from the date of the note?	0	0
9. What is the square root of 5/9 of 4/5 of 4/7 of 7/9?	174	56.5
10. The city of Boston has 120,000 inhabitants, half males, and its property liable to taxation is one hundred millions. It levies a poll of 2/3 of a dollar each of one half of its male population. It taxes income to the amount of $50,000, and its whole tax is $770,000. What should a man pay whose taxable property amounts to $190,000?	1	0.3

The North American Arithmetic: Part Third, for Advanced Scholars (Caldwell & Courtis, 1925). Allegations of poor performance by Mann (1845) and his committee therefore represented a major challenge to the cyphering approach, itself, and to the masters and "ushers" (the name give to support teachers) who had supervised the students' work. The results were widely publicized (Katz, 1968).

Mann maintained that the results showed that the cyphering approach was not up to the challenge of providing valid assessment of student, and therefore teacher, efficiency. That raised the question of how arithmetic was best taught, and those responsible for administering the recently-established normal schools believed they had an answer to that question.

The externally-set written examinations that Mann and his committee imposed upon the Boston teachers and students were not the first of this kind to be used in the United States. Thus, for example, the first issue of the Pennsylvania *Common School Journal* (Volume 1, number 1, published in 1844) reproduced copies of written examinations (including arithmetic and algebra examinations) used for the selection of new Pennsylvania public school teachers. But Mann's use of written examinations as a weapon in his dispute with the Boston schoolmasters received much publicity, and served to legitimise the new form of evaluation and assessment.

Those who supported the introduction of externally-set written examinations did not anticipate some of the weaknesses of that system of assessment (Katz, 1968; White, 1886). Such was the influence of those who supported this approach that soon it was adopted in many states (Landis, 1853; Kilpatrick, 1992; Reisner, 1930; S. H. M., 1856). Under the new system, the assessment of a teacher's worth was no longer intimately related to the quality of students' cyphering books and that hastened the demise of cyphering books in schools. Teachers began to think that they could not afford to allow their students to spend time preparing cyphering books, because they needed as much time as possible to prepare students for the forthcoming high-stakes written examinations.

Opposition to the cyphering-book tradition in normal schools. During the late 18th century and throughout all of the 19th century, "normal schools," at which prospective teachers were trained, became increasingly popular in Europe (Harper, 1935, 1939). That was also true in North America after 1839—the year when the first public normal school in the United States was established. Among other things, the normal schools would teach generations of prospective and practicing teachers that the inductive methods advocated by Pestalozzi and Colburn, were needed in schools (Barnard, 1851, 1856, 1859; Monroe, 1969), and that, from a quality-of-learning perspective, the cyphering tradition had produced unsatisfactory results (Harper, 1935, 1939; Henry, 1843; Page, 1877; Reisner, 1930; Wayland, 1842). Normal school graduates were expected to have acquired a sufficiently strong knowledge of mathematics that they could use whole-class teaching methods when teaching arithmetic.

The views on cyphering of Nicholas Tillinghast (1804–1855) and Richard Edwards (1822–1908), two noteworthy principals of the early normal school movement, are especially worthy of consideration. Tillinghast, chosen in 1841 by Horace Mann to be the foundation Principal of Bridgewater Normal School, in Massachusetts, held that position until his death in 1857. Richard Edwards, one of Tillinghast's Bridgewater students, was Principal of Illinois State Normal University (ISNU) between 1862 and 1876 after having been principal of two other normal schools before that. Tillinghast was a scholar who had been trained in mathematics at the West Point Military Academy, which followed a graded-class approach to mathematics education.

While at Bridgewater, Tillinghast wrote two books, *Elements of Plane Geometry for the Use of Schools* (1844) and *Prayers for the Use of Schools* (1852). According to Edwards (1857), the titles of those books summed up the kind of person Tillinghast was—simultaneously, he was a deeply spiritual person as well as an intellectual leader. According to Dana P. Colburn (1825–1860), a student at Bridgwater State Normal School under Tillinghast and later an author of arithmetics and Principal of Rhode Island State Normal School, he was more indebted for ideas on arithmetic to Tillinghast than to any other person. Colburn (1855) was moved to write: "Those who best know him, will best understand how great that indebtedness must be" (p. iii). Stimulated by leaders like Tillinghast, Edwards, and Colburn, the North American normal schools trained generations of public school teachers (Boyden, 1933) to despise the old cyphering approach to arithmetic.

Tillinghast was imbued with the spirit of Pestalozzi and believed that all branches of mathematics deserved to be taught as vibrant but rigorous subjects. For him, the chief aim was for students to *understand* what they learned. Richard Edwards (1857) said of Tillinghast:

> There was thoroughness in his teaching, but there was also another element, which if we could coin a word we might call "logicalness"—an arranging of the subject taught according to the character and wants of the mind to be instructed. In every operation, there was not only thorough knowledge, but also thorough reasoning. Every point was not only to be thoroughly understood, but it was to be understood rationally not only by itself, but also in its relations. The pupil was himself required to discover if possible, or at least to appreciate, the connection between one part of the subject and another, to see how much of one statement could be inferred from a previous one. More thoroughness in the knowledge of facts, or of principles, learned and remembered, is a very different matter from the thoroughness that characterized the teaching of Mr Tillinghast. The one can be accomplished by the industry of the pupil; the other requires, in addition, careful thought and ready skill on the part of the teacher. (p. 14)

This emphasis on the need for teachers to be active in helping students understand the principles of written arithmetic had been noted in 1847 by David Page, Principal of the State Normal School at Albany, in New York. Page (1877) unambiguously referred to the cyphering approach as "the old plan" (p. 53), and said that he himself had not understood the principles of arithmetic after he had "ciphered through" some four or five arithmetics (p. 53).

Page was probably the main author of a "Report of the Executive Committee of the State Normal School" for New York that appeared in the *District School Journal of the State of New York* in April 1846, and therefore comments on the teaching of arithmetic, in that report, are of special interest. The *Report* summarized the Normal School's approach to arithmetic instruction:

> In commencing the mathematical course, it was thought that *thoroughness* alone could secure a pleasant and profitable progress. To gain this, instruction commenced at the fundamental principles of arithmetic. The students were required to solve *orally* and without the aid of a book, all the questions in "Colburn's Intellectual Arithmetic." After the attainment of considerable proficiency in this exercise, they were allowed to propose to each other

such questions as involved the principles already acquired. This gave additional interest to the subject of study; while the brevity and clearness displayed in stating the questions, and the facility and ingenuity in solving them, clearly proved that the students were making not only a thorough but rapid improvement. (p. 2)

There was no mention of cyphering in either the above or the following passage:

In teaching written arithmetic, great care was taken that the principles on which the rules were grounded, should be fully comprehended. To this end, the pupils were required to go to the black board, and taking the position of a teacher, to go carefully through the analysis of each topic; while any member of the class was permitted to point out whatever he deemed incorrect or defective, and the *temporary* teacher was called on to defend his course, or to correct his mistake. Thus rigid criticism was encouraged, and no subject was dismissed, until it was so well understood, that any of the class could act the part of a teacher, and explain it at the black board. Frequently, several members of the class were called on in succession to elucidate the same subject; thus affording an opportunity for comparing the relative merits of various methods. (p. 2, original emphasis retained)

This whole-class approach, which involved pupils in taking the role of the teacher at the blackboard (Holbrook, 1859), contrasted with the old cyphering approach. But, the new approach could only be adopted in graded schools in which classrooms had blackboards. Although blackboards had been introduced into some American schools in the 1820s and 1830s (Kidwell et al., 2008), in the 1840s most North American schoolchildren still did not attend schools which had blackboards. It was also true that in the 1840s most school students learned arithmetic in ungraded classrooms (Finkelstein, 1989; Harper, 1935, 1939).

Early normal school leaders, like Dana Colburn, Richard Edwards, Alfred Holbrook, David Page, Cyrus Peirce and Nicolas Tillinghast decried the cyphering tradition's emphasis on rules and cases, and on copying (Katz, 1968; Reisner, 1930). Page (1877) maintained that a teacher of arithmetic should understand the subject so well that "he could teach it thoroughly though all text-books should be excluded from his school-room" (p. 54).

Henry Barnard (1856), superintendent of common schools and principal of the Connecticut State Normal School between 1851 and 1855, was adamant that the old ungraded system was hopelessly inefficient. Barnard believed that most teachers in ungraded classrooms were simply unable to attend to the individual needs of so many children with so many different levels of language, social development, numerical skill, and knowledge of curriculum content (Cajori, 1890; Katz, 1968). This often resulted, Barnard (1851, 1856) maintained, in teachers feeling the need to enforce draconian systems of discipline (Richeson, 1935).

These normal-school reformers set out to create a new breed of teachers who would invite their students to explore, discover, and understand mathematical relationships. This was to be achieved through a combination of whole-class instruction to groups of students, the use of quality textbooks with well-graded problems, individual assessment through recitation, and well-constructed written tests. Whole-class instruction, in particular, was regarded as an essential characteristic of reform methods, and was supposed to feature challenging questions in which students were expected to reflect and reason. From the normal-school perspective, the cyphering approach was abhorrent, something totally antithetical to sound education philosophy.

Richard Edwards was originally appointed to Illinois State Normal University (ISNU) as head of its mathematics department, but on assuming the principalship he arranged for another Bridgewater/Tillinghast graduate, Thomas Metcalf, to come to ISNU as head of the mathematics department (Marshall, 1956). Metcalf would remain at ISNU for over 30 years. With leaders like Tillinghast, Edwards, Metcalf, Henry Barnard, Edward Brooks, Horace Mann (Mann would later became President of Antioch College, a normal school in Ohio), David Page (1877) and Cyrus Peirce (Harper, 1939), the normal schools responded to the challenge of changing the intended, implemented and attained curricula of North American schools, at all levels (Barnard, 1851, 1856, 1859). Normal school students were taught that successful mathematics teaching and assessment not only required careful verbal questioning of what students knew, but also of how and why they knew it, and how and why it might be useful (Brooks, 1860; Edwards, 1857; Executive Committee of the State Normal School, New York, 1846; Thayer, 1928). And that was something which, the normal school leaders believed, had *never* been achieved through the cyphering approach (State of Massachusetts, 1855).

The use of written examinations quickly became common-place in the normal schools, and this meant that a cadre of teachers was trained for whom the evaluative function of cyphering books was rejected as inappropriate (Barnard, 1851; Reisner, 1930; White, 1886). The twin message of teaching for understanding through active student engagement and assessing learning through written examinations was conveyed in all normal schools and at summer institutes that provided professional development for practicing teachers. The determination to convey that message, and implications of new policy whereby results from externally-set written examinations would provide the main criteria for measuring student and teacher efficiency, marked the beginning of the end for the cyphering-book tradition in North America.

Many faculty and graduates of normal schools wrote mathematics textbooks. On that score, the lengthy honor roll included Robert F. Anderson, M. A. Bailey, Howard Griffith Burdge, Edward Brooks, Dana P. Colburn, John W. Cook, Charles Davies, James B. Dodd, David Felmley, S. A. Felter, Benjamin Greenleaf, Daniel B. Hagar, W. D. Henkel, Alfred Holbrook, George W. Hull, Edwin C. Hewett, Malcolm MacVicar, Horace Mann, Charles A. McMurry, Frank M. McMurry, William J. Milne, George Perkins, Albert N. Raub, Martha H. Rodgers, John Herbert Sangster, David M. Sensenig, G. C. Shutts, David Eugene Smith, L. M. Sniff, John F. Stoddard, Nicholas Tillinghast, Electa N. Walton, and George A. Walton.

Whether all of these educators were well qualified to write mathematics textbooks is a moot point, as is the extent to which the changes in school mathematics which they sought were educationally sound (see, e.g., Boston Schoolmasters, 1844; Richeson, 1935; Smith, 1973) or ever actually achieved. But, on key curriculum, teaching, and assessment issues, their voices were heard, and from a mathematics education perspective they represented the new broom.

The move towards graded schools. Historians (e.g., Charles Harper, 1935, 1939; Michael Katz, 1968) have tended to point to the work in the 1830s and 1840s of Albert Picket, in Ohio, and Cyrus Peirce and John D. Philbrick, in Massachusetts, as leaders of the revolution in school management that culminated in the widespread acceptance of what came to be known as "grade" (or "graded") schools in the United States. These reformers called for curricula that were carefully sequenced according to complexity, the aim being to allow pupils to be grouped together. By this new system, larger schools in cities such as Boston, New York, and Philadelphia were organized around seven or eight different levels, with within-subject sequencing based largely on complexity, and only partly on age. Students who were at about the same stage in their academic progress could study together in the one room, and this, it was argued, made discipline easier to enforce because it facilitated an orderly development of subject knowledge (Watras, 2002). Furthermore, teachers could be given teaching assignments that matched their training and interests.

Following the example of colleges like West Point Military Academy (Hoskin, 1994), most larger cities had, by 1860, moved to graded public and high schools, and schools were being built that had more rooms than had been the case 30 years earlier (Katz, 1968; Tyack & Tobin, 1994; White, 1886). This transition was often vehemently resisted by experienced teachers in the old academies (see, e.g., Boston schoolmasters, 1844). Other critics, like James B. Dodd (1856), of Transylvania University, recognized a danger in the almost universal adoption of the blackboard and oral system, particularly if the teachers did not know their mathematics sufficiently well. Dodd wrote of "noise and confusion" (p. 13), and stated that in the new state of affairs "there was too much of teaching and too little of studying in many schools" (p. 13). However, all over the nation, whole-class teaching was enthusiastically endorsed and implemented by carefully appointed state superintendents and local education officials.

Leadership within the schools, at both the elementary and high school levels, came from young teachers who had been trained in normal schools. In arithmetic classes, teachers were now expected to give whole-class lessons based on textbooks. Students had personal copies of textbooks and were expected to prepare for the next period of recitation, or written test, by answering exercises set in those textbooks (Angus, Mirel, & Vinovskis, 1988; Watras, 2002).

Between 1840 and 1860 the die was cast against the cyphering tradition, and after 1860 the tradition quickly and quietly became a thing of the past. Many teachers had, themselves, found it difficult to learn arithmetic well under the old system and school reformers were confident that the new state of affairs represented a change for the better (Dodd, 1856; Hummel, 1964). They maintained that the old curriculum, with its strong commercial orientation, and the old pedagogy, with its emphasis on memorization, were increasingly ill-suited to an age when all students in common schools would be expected to spend a lot of time trying to learn arithmetic. As William Shoup (1889), a school principal in Iowa in the 1880s, would write:

> Complaint is frequently heard that the education given by the common schools is not practical. ... Can your scholars who are struggling with equation of payments, averaging accounts, foreign exchange, custom-house business, and other matters with which they are

about as liable to have dealings in after life as with building a railroad to the moon, add a column of figures with reasonable rapidity and accuracy? If not, there is some foundation for the complaint that the schools are not as practical as they ought to be. (p. 141)

In 1800, the *abbaco* curriculum may have been fine for children in schools in New England's trading centers, but by 1850 many educators around the nation had come to believe that it was not suited to an age when the principle of arithmetic-for-all needed to be accepted by educators, parents, and employers. Some astute commentators were concerned, however, that the new régime, based on whole-class teaching, would result in too much lecturing and not enough learning, and were worried that benefits arising from recitations would be lost (Orcutt, 1859; Richeson, 1935; Smith, 1973).

Normal Schools, Public High Schools, New Textbook Series, and Cyphering Books

The movement against the cyphering tradition occurred at the same time as the establishment and growth of district high schools (Inglis, 1911; U.S. Department of Commerce, 1975; Vinovskis, 1985), and also at the same time as a sharp rise in the number of home-grown mathematics textbooks (Karpinski, 1980; Michalowicz & Howard, 2003). Given that many high school students studied arithmetic, the decline in the usage of cyphering books could be regarded as a phenomenon brought about by high-school opposition to cyphering books.

During Horace Mann's quarrel with the Boston schoolmasters in the mid-1840s the schoolmasters alleged that Mann had not had enough experience as a classroom teacher to know what was best so far as teaching methods were concerned (Boston schoolmasters, 1844; Caldwell & Courtis, 1925; Katz, 1968). Mann denied the charge, and countered by encouraging normal schools, which were seen by many as his brain-child, to send their most highly trained and obviously talented graduates into the new high schools in and around Massachusetts (Barnard, 1851; Boyden, 1933; Edwards, 1857; Harper, 1935; Reisner, 1930). A similar series of events would occur in other states (see, e.g., Barnard, 1851; Page, 1877). The new high school teachers did not ask students to prepare cyphering books for they had been trained to believe that they knew of a more excellent way to develop arithmetical knowledge and skills (Massachusetts Board of Education, 1855). This was one of the causes of continuing friction between older teachers and the "new" practitioners arriving from the normal schools (Edwards, 1857).

Eileen Donoghue (2003b) has argued that "the field of mathematics education in the United States traces its origins to the 1890s" (p. 159), and has linked that to the rise of American research universities. It is true that there was no professional association of mathematics teachers in the United States during the 19th century, but during the period 1840–1890 there was much reflection on the role of the normal schools, and on curriculum, pedagogical and assessment issues in relation to school mathematics (see, for example, Boston Schoolmasters, 1844; Dodd, 1856;

Massachusetts Board of Education, 1855). In the same period, large numbers of teachers attended teacher institutes during the summer (Harper, 1935, 1939).

Consider, for example, the work of Edward Brooks. In 1855, Brooks accepted a position at the State Normal School at Millersville, Pennsylvania—at that time, John F. Stoddard, author of numerous school mathematics texts, was Principal—and later Brooks served as Principal and Professor of Mathematics at the Millersville Normal School before becoming Superintendent of Public Schools of Philadelphia. Between 1858 and 1900, Brooks authored a popular series of school mathematics textbooks (e.g., Brooks, 1860). He also wrote *Philosophy of Arithmetic as Developed from the Three Fundamental Processes of Synthesis, Analysis, and Comparison Containing also a History of Arithmetic* (Brooks, 1876) and *Normal Methods of Teaching Containing a Brief Statement of the Principles and Methods of the Science and Art of Teaching* (Brooks, 1879). The term "mathematics education" was not used then, but the very titles of these books suggest that Brooks could be regarded as a pioneer in the field of mathematics education.

During the 1840s there was a sharp rise in the number of mathematics textbooks written for schools in the United States. At that time, series of texts written by Charles Davies, Frederick Emerson, Benjamin Greenleaf, Stephen Pike, Joseph Ray, and John Stoddard were published and these competed with well-established textbooks by Warren Colburn, Nathan Daboll, and Daniel Adams. We believe, however, that the logic of linking the greater availability of textbooks with the decline in the usage of cyphering books is spurious. A counter-example is provided by noting that during the period 1800–1830 there had been a similar rise in the availability of textbooks (at that time, the most popular school arithmetics were those written by prolific authors like Nicolas Pike, Daniel Adams, Warren Colburn, Nathan Daboll, Zachariah Jess, Stephen Pike and Michael Walsh), and yet there had been no sharp decline in the usage of cyphering books.

Chapter 7
Conclusions, and Some Final Comments

Abstract The five research questions identified in Chapter 4 are answered in this chapter. It is argued that the cyphering tradition inculcated among students an "ownership" of the mathematics they studied, and that many students who prepared cyphering books proudly used these as reference books for the rest of their lives. The chapter closes with a list of researchable questions, associated with the cyphering tradition, for future investigation.

Answers to Research Questions

In Chapter 4 of this book we described the circumstances by which we arrived at the following five key research questions relating to the 212 manuscripts in our PDS.

1. Where and when were the cyphering books prepared, and by whom?
2. To what extent were the manuscripts consistent with *abbaco traditions*, especially in relation to (a) mathematical content and sequencing; (b) genre; and (c) writing-as-also-arithmeticke considerations?
3. What theoretical base, if any, can be identified to encapsulate the educational purposes of the cyphering books?
4. Why, between 1840 and 1860, was there such a sharp decline in the use of the cyphering approach in the Unites States?
5. From an education learning perspective, what were the advantages and disadvantages of the cyphering approach?

We now answer each of those questions.

Research Question 1

Where and when were the cyphering books prepared, and by whom? Answers to these questions were suggested by entries in Table 5.1 and by Figure 5.1, and in the ensuing discussion (in Chapter 5). There, we reported that of the 189 manuscripts in the PDS on which locations of writers were stated, 88 were prepared in a New England state, 39 in Pennsylvania, and 43 in New York, New Jersey, or Maryland.

We do not mean to imply that the preparation of cyphering books was confined to these states or regions: we know, for example, that there are about 60 cyphering books held in the Southern Historical Collection at the University of North Carolina at Chapel Hill (Doar, 2006) that originated from other regions.

The manuscripts in our collection are largely silent on the conditions under which they were prepared. We do not know how many were prepared in local subscription schools, how many in evening schools for apprentices, how many in schools conducted by private reckoning or writing masters, or how many by students working alone, from a textbook or another cyphering book. That suggests a further investigation: Under what conditions were they prepared?

Data summarized in Figures 5.1 and 5.2, together with data gleaned from manuscripts held in the Phillips Library, the Houghton Library, the Clements Library, the David Eugene Smith Collection, the Wilson Library at the University of North Carolina at Chapel Hill, the Swem Library at the College of William and Mary, the New York Public Library, and the Huguenot Historical Society in New Paltz, New York, provide compelling evidence that cyphering was the dominant instructional method in mathematics throughout the period 1700–1830. Nevertheless, any inferences made about the cyphering situation in 18th-century North America should take into account the fact that most of the original cyphering books no longer exist. That lesson comes from a consideration of hornbooks. We know that hornbooks were widely used in the North American colonies and yet very few of them are extant, and almost all of those that remain were imported from Europe before they were used in North America (Earle, 1899; Plimpton, 1916).

The role of cyphering books in the period 1830–1860 is also difficult to interpret because of the growing population, and also because of the possibility that as a result of Warren Colburn's emphasis on inductive reasoning many more young children (under 10 years of age) than ever before learned arithmetic in schools. If this were indeed the case then a higher percentage of students than ever before should have been "ready" to begin cyphering. An initial reaction to the data, however, certainly suggests that the cyphering-book tradition remained alive in the 1830s. The period 1840–1865 saw its demise.

The cyphering books that we hold that are not within the PDS include 42 prepared in England and one in Scotland. Five of these manuscripts were prepared after 1861—in fact, two were prepared in the 1860s, two in the 1870s, and one in the 1880s. This suggests that the cyphering-book tradition lingered for longer in Great Britain than it did in North America. Whether that was indeed the case and, if so, why, are matters for further research.

Research Question 2

To what extent were the manuscripts consistent with *abbaco traditions*? We shall consider that question in relation to (a) mathematical content and sequencing; (b) genre; and (c) writing-as-also-arithmeticke considerations.

Mathematical content and sequencing. The content in the PDS cyphering books is remarkably consistent with that in the *abbaci*, suggesting that the cyphering tradition in North American was a late flowering of a 600-year tradition. With cyphering books that focused on elementary arithmetic, early topics were notation and numeration according to Hindu-Arabic place-value notation, followed by the four operations on whole numbers and denominate quantities. Then followed currency conversions, reduction, practice, and vulgar and decimal fractions. One or two of the various rules of three would often mark the end of the elementary content stage, or the beginning of a more advanced content stage.

Middle-level arithmetic content included a wide range of business topics, including loss and gain, barter, simple and compound interest, fellowship, alligation, false position, annuities, and equation of payments. For the small number of students who went further, topics like involution and evolution, permutations and combinations, arithmetical and geometrical progressions and, perhaps, logarithms were covered.

There were surprisingly few cyphering books in which the focus was on algebra or geometry. A standard sequence of topics featured in the algebra manuscripts, and Euclidean straight-edge and compass constructions provided the main content in the geometry manuscripts. It is also the case that there were fewer manuscripts that dealt with trigonometry, or navigation, or surveying than we expected, but it was in these manuscripts that the highest level of calligraphy and the most intricate and beautifully drawn diagrams appeared. Our analyses of the collections of cyphering books at the Phillips Library (Peabody Essex Museum, Salem) the Houghton Library (Harvard University), the Clements Library (the University of Michigan), and Ashley Doar's (2006) analyses of the manuscripts held in the Southern Historical Collection at the University of North Carolina have indicated that those collections include a higher proportion of manuscripts dealing with navigation or surveying than we have in our PDS.

Genre. Although the definition, rule, case, type problem, exercise sequence that Van Egmond (1980) found in the early European *abbaci* was an important aspect of the content of most the cyphering books in the PDS, the rhetorical genre of the early *abbaci* that Van Egmond reported was not. Solutions to both model problems and exercises tended to be set out in sequences of applications of algorithms, with answers written immediately after the final algorithm had been applied. Compared with the reflective, discursive genre of earlier *abbaci*, the genre in the cyphering books in our PDS was mechanical. Equals signs and logical connectives were rarely used. It would be an interesting research exercise to trace the change from the reflective, discursive genre of early *abbaci* to the algorithmic genre that is so common in schools of today.

Writing-as-also-arithmeticke considerations. The tradition that cyphering books should feature high-level penmanship and calligraphy, and high quality diagrams (Carpenter, 1963; Monaghan, 2007)—if illustrations were necessary—was still being followed by most of those who prepared the books in our PDS. There is an obvious difference in the standard of presentation in our PDS cyphering books and in the exercise books in which arithmetic or other forms of mathematics were written after the 1860s. The fact that many writing masters doubled as teachers of arithmetic—for example, even the renowned British mathematician, Charles Hutton, once described himself as "writing master and teacher of mathematics" (Hutton, 1766, p. i)—led to pressure being placed on students to ensure that their best penmanship and calligraphy featured in their cyphering books.

That said, when we examined the penmanship, calligraphy, and illustrations in the 43 cyphering books from Great Britain in our "extra" collection, we were struck by the relative calligraphic excellence displayed in the British manuscripts. We would conjecture that the long traditions established by Christ's Hospital (Reynolds, 1818), and by writing masters such as George Bickham, Samuel Butler, Edward Cocker, and James Hodder, resulted in the forms of penmanship and calligraphy in cyphering books prepared in Great Britain being different from those prepared in North America. That conjecture suggests a useful agenda for further research.

Research Question 3

What theoretical base, if any, can be identified to encapsulate the educational purposes of the cyphering books? Warren Van Egmond (1980) mentioned that when he began studying *abbaci* he could not identify any underlying, implicit rationale or theory of education. Later, he began to see the *abbaci* as structurally quite similar. We have had a parallel experience. After studying the contents of the manuscripts in our PDS, we identified and reported an educational rationale on which they were based (Ellerton & Clements, 2009b).

Our analyses of North American cyphering books suggested that, as with the earlier European teachers of *abbaci*, the fundamental aim of most teachers of arithmetic and reckoning masters whose students prepared cyphering books was to assist them to become independent problem solvers, for life. This point of view is consistent with that put forward by Albrecht Heeffer (2008) when discussing the genre of algebra *maestri d'abbaco*. Heeffer stated that practically every text dealing with algebra featured the same structure, which could be divided into six parts: problem enunciation, choice of rhetorical unknown, manipulation of polynomials, reduction to a canonical form, applying a rule, and numerical test. Although IRCEE—the acronym which we have used to represent our summary of the main genre to be found in arithmetic *abbacus* manuscripts—is different from this list, there are striking similarities.

We believe that students were not usually aware that they were being invited to recognize important structural similarities in carefully chosen, mathematics-based, problem scenarios, and to select and apply appropriate rules to solve the problems. However, most of the major problem categories had sub-categories, and for each

Answers to Research Questions 143

sub-category a type problem was chosen and solved. Students were then invited, individually, to tackle problems that were structurally identical to the type problems. Before they were permitted to enter solutions in their cyphering books they had to be checked for correctness. When, finally, solutions were handwritten into cyphering books, it was expected that students displayed excellent penmanship.

After repeating this sequence for all the rules and cases associated with several main categories of problems, students were asked to solve miscellaneous problems (or "promiscuous questions") that embodied the structures that had just been studied. Each was concerned with a real-life situation that was either meaningful to the student then, or might be meaningful in the not-too-distant future. Each student was expected to recognize the mathematical structure, solve the problem, have the solution checked, and write the solution in his or her cyphering book. It was assumed that writing correct solutions to problems compelled students to "reflect" on problem structures, and made them likely to recognize and solve structurally similar problems in the future (Van Egmond, 1980). Although this theoretical rationale, which is summarized in Figure 7.1, was recognized by the best teachers and students, it was never explicitly recorded.

The rationale might be regarded as defining a "socially-oriented, structure-based, problem-solving" theory of learning. It was "socially-oriented" because

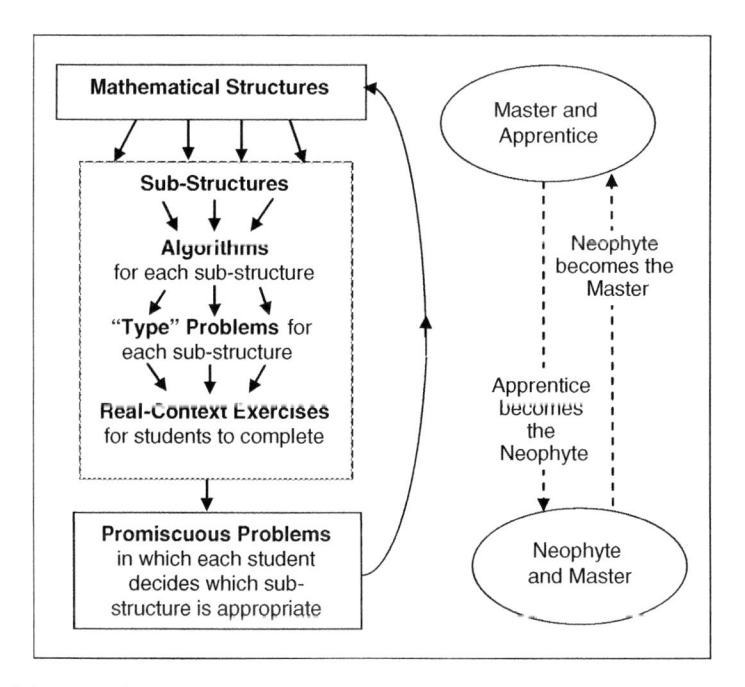

Figure 7.1. Educational rationale for the cyphering approach (Ellerton & Clements, 2009b).

type problems were deliberately chosen by instructors so that they were likely to be relevant to present or future life situations of the students. It was "structure-based" because each problem was chosen because it offered the opportunity to help the student to recognize that a given problem needed mathematics of a particular kind, with a special structure. It was "problem-solving" because students were expected to learn to solve problems like those under consideration.

A sense of apprenticeship theory was incorporated into the model, with masters assisting students to learn to solve mathematics problems efficiently. The theory also incorporated the idea that it was not enough for students to be able to apply algorithms to solve problems—they needed to demonstrate, repeatedly, that they would be able to work out which mathematics was needed when they were confronted with unseen ("promiscuous") problems for which the structures were not immediately obvious.

After we developed the theoretical perspective illustrated in Figure 7.1, we found a similar perspective outlined by Danny J. Beckers (1999) for cyphering books in the Netherlands between 1750 and 1850. It is intriguing that D. Senthil Babu (2007) identified a similar structure in cyphering books prepared in Tinnai schools in Tamil Nadu, India, in the 18th and 19th centuries. Babu identified the following steps in the problem-solving process that Tamil boys were expected to go through when attempting to solve a given problem:

1. Recognize the variable in the problem;
2. Identify the appropriate algorithm, or sequence of algorithms;
3. Recollect how the relevant algorithm(s) should be applied;
4. Apply the algorithm(s) to solving the problem;
5. Add the new problem type to the already-existing repository of skills;
6. Conceptualize the new class of problems within the overall system of algorithms and known problem types.

Babu (2007) maintained that from a mathematics education point of view, "skill competence and functionality, with a thorough orientation to the local society, marked the curriculum of these [Tinnai] schools" (p. 15). Memory was the modality of learning in the schools, but students learned to cope with the calculational needs of their societies. Furthermore, those who wished to go on to higher learning were also well prepared by their training in problem posing and problem solving.

An interesting question was whether this type of structure was to be found wherever cyphering manuscripts were prepared across the world in the 18th and 19th centuries. After all, in the late 12th century, Leonardo Pisarno identified the approach among Arab nations in Africa, and tradition has it that it was through his *Liber Abbaci* that the tradition made its way into Europe and from there to America. We know that the Hindu-Arabic numeral system featured in India before that, and it is possible that what Babu identified in the Tinnai schools was a tradition that existed *before* the European colonizers came to India.

Research Question 4

Why, between 1840 and 1860, was there such a sharp decline in the use of the cyphering approach in the Unites States? At the outset, we should say that we wondered how a 600-year tradition, which was still widely appreciated in the 1830s, could lose its influence so quickly. We considered the possible explanation that the potentially transformative efforts of Warren Colburn in the 1820s led to the abandonment of the cyphering approach, but rejected it because cyphering books were still being widely used in North America in the 1830s and into the 1840s. We also rejected the related possible explanation that the advent of new textbook series, written by prolific authors such as Warren Colburn, Charles Davies, Joseph Ray and Benjamin Greenleaf, pushed the cyphering approach out of the schools. We rejected that thesis because during the period 1800–1830 the cyphering-book tradition continued to exert a powerful influence despite the ever-increasing popularity of books written by earlier authors like Nicolas Pike, Daniel Adams, Nathan Daboll, Zachariah Jess, Michael Walsh, and Stephen Pike.

After weighing up the variables that could have singly, or in combination, contributed to the demise, we identified three major contributing factors. The first was the introduction of written examinations in the 1840s. In the past, committee and parent perceptions of the quality of cyphering books had played a key role in the evaluation of the work of students and teachers, and this had meant, for both teachers and students, that time spent on cyphering books was likely to be rewarded. This state of affairs changed abruptly with the introduction of written examinations, and the subsequent public reporting of school results. Now teachers wondered whether they should abandon the cyphering tradition, given that preparation of high quality cyphering books would not necessarily help them get a teaching position in the future. In the new state of affairs, they needed more time to prepare students for written tests, and any time that students spent on preparing cyphering books was regarded as "unproductive."

In the United States early in the 21st century, many believed that the "No Child Left Behind" legislation, which calls for links to be made between performances on written examinations and the quality of student learning, teacher instruction, and school organization, had an effect on what transpired in mathematics classrooms (Dossey, Halvorsen, & McCrone, 2008; Wright, Wright, & Heath, 2008). We believe a similar phenomenon occurred in the 1840s and 1850s, and contributed to the rapid demise of the cyphering approach in schools (White, 1886).

A second major force that precipitated and hastened the demise of the cyphering approach was the extreme distaste for that approach among leaders of the recently-established normal schools. Graded schools were becoming the educational fashion of the age, and although these were not really appropriate in small one-room schoolhouses, where most of the nation's children still went to school, the whole-class methodologies were pushed strongly in normal schools, and were buttressed by the attitudes of state education officials like Horace Mann and Henry Barnard. Normal school principals and district education officers wanted a whole-class approach to teaching arithmetic, complemented by group recitations (Thayer, 1928). Inspired by

Pestalozzian inductive theory, they trained several new generations of teachers, the best of whom went directly into the high schools, to apply what they had learned in their professional training.

Graduates from the best normal schools learned to believe that a new whole-class teaching approach should replace the more impersonal, rules-based, individual approach that had historically been such a vital component of the cyphering tradition. Normal-school students were trained to believe that the cyphering approach to teaching arithmetic was discredited. When they graduated and became full-time, fully qualified teachers they attempted to adopt interactive whole-class methods which, they expected, would result in students gaining deeper, more connected understandings of the subject.

The large influence of normal schools on teaching practices was not confined to North America. In Japan, for example, Masami Isoda (2007) has shown that large-scale changes in mathematics teaching approaches occurred in Japan around 1870. At that time, the individual, "samurai-type" learning environments, that had characterized school education in the first 70 years of the 19th century in Japan, gave way to more interactive, inductive, teacher-led approaches. According to Isoda, the main force influencing change was the adoption by young teachers of pedagogies recommended at a normal school established in Tokyo around that time.

The third major force that hastened the demise of the cyphering approach was an increasing recognition that an *abbaco* arithmetic curriculum was not suited to the needs of American society. At a time when much larger numbers of North American children were attending school and studying arithmetic than ever before, the old emphasis on business-related arithmetic was seen to be losing relevance, especially for students who did not live in or around mercantile centers. School authorities declared that they wanted students to spend more time on algebra, and university and college authorities were demanding that those seeking admission to their courses should be able to demonstrate knowledge of algebra and geometry. If girls as well as boys would study arithmetic, then the old focus on mercantile arithmetic simply had to be abandoned.

Research Question 5

From a learning perspective, what were the advantages and disadvantages of the cyphering approach? Even a cursory examination of the cyphering books in the PDS would suggest that many, and probably most, of the individual writers who prepared them were immensely proud of their books. However, Florence Yeldham (1936) claimed that in Great Britain students grew to view their cyphering books as "distasteful" (p. 130), and certainly Daniel Dowling (1829) thought that the cyphering tradition had "crowded [students] with gross absurdities, unintelligible rules, and useless examples" (p. vi).

How did the North American students who prepared the cyphering books in the PDS regard their experiences in preparing the manuscript? Although we might think that the manuscripts themselves are silent with respect to that question, it could be argued that they testify to energy, correctness, and care—care by students in getting and writing down correct solutions, and care by teachers not only in choosing appropriate exercises but also in ensuring that student solutions were correct. The cyphering tradition incorporated an individualized approach to mathematics education, with each student tackling problems that his or her teacher deemed he or she was ready to do. Students were expected to "get right answers" before entering solutions in their cyphering books, and this required the teacher to check individual students' "rough solutions" before final solutions were written in cyphering books. The whole-class approach to teaching arithmetic which replaced the cyphering approach was much less likely to be associated with such a high degree of individualization and teacher attention to the progress of students.

Some critics have maintained that with the cyphering approach it was unusual for teachers to speak with their pupils about arithmetic (or whatever branch of mathematics was the main focus). But William Butler (1788), after stating that it was "of the utmost importance to have the pupils grounded in the several tables," went on to say: "It is necessary to make them, on presenting a sum which they have worked, define the rule to which it belongs, and assign a reason for every part of the operation, until they have attained a thorough knowledge of the whole process" (p. iii). According to Butler, the teacher needed to question the student about meaning, and about how parts of a problem related to each other, and about reasons for rules. Butler (1806) emphasized that teachers should talk regularly with individual students about mathematics. Vivian Thayer (1928) quoted Henry K. Oliver, a pupil in Boston around 1800, as stating that every school day he received individualized instruction, at his teacher's desk, for 40 min, and that for the rest of the day he was expected to work on what was set by the teacher.

There is evidence that cyphering books were used by their owners long after they had prepared them. An 1823–1824 cyphering book originally prepared by an 18-year-old Ezra Cornell provides a case in point. The book is held in Ithaca, New York, at the University that now bears the owner's name. Later in his life, Cornell (1860) commented that at the time he prepared his cyphering book he was a "young boy from the farm." In his book he wrote practical rules and exercises in arithmetic, beginning with the four operations and proceeding to vulgar fractions, simple and compound interest, measurements, equation of payments, currency conversions, annuities, and barter. Toward the end of the book, in 1824, his topic was "loss and gain." Around 1860, when he was in his mid-50s, Cornell made several annotations to "his book." Next to his 1824 entries on "loss and gain" he wrote:

> The above, or the examples in the above rules, though not entered in this book I have been practicing ever since the above was written, and it has been a life long struggle to see which would preponderate. In 1844, there was a balance of perhaps of a couple of thousand dollars on the credit side. In 1854 the contest was a doubtful one, and a debt with which I was then incumbered (sic.) amounting to $50,000 would probably have swept the board if the game had been stopped at that period. But the contest has been continued, with increasing success for the side of gain and at the present period Feb. 1, 1860, that mountain of debt has mostly been paid at the rate of 100 cents in a dollar with 7 per cent interest added, and a yearly income of $15,000, seems to be a reliable guarenty (sic.) that the credit side has won the victory. Ezra Cornell, Forest Park, Ithaca, Tompkins Co. N.Y., February 1, 1860

History is largely silent on how many others made direct use, in their later-lives, of cyphering books that they prepared when they were young. But one can imagine that there are many others with stories like Ezra Cornell's. George Littlefield (1904) has claimed, for example, that the book from which a young George Washington learned to cypher "undoubtedly exerted a tremendous influence" on the future President's subsequent life (p. 194).

What were some of the *disadvantages* of the cyphering approach as it was practiced in North America? One only has to look at some of the writings of people like Reuben Chambers (1835) and Lyman Cobb (1835) to begin to realize that sometimes teachers regarded the quality of the penmanship and calligraphy as far more important than the actual arithmetic itself. Although that is understandable, given the committee system of evaluating the worth of teachers, the result was that for many students, and perhaps most students, more attention was paid to form rather than to substance (Page, 1877).

It could be argued that although the extraordinarily beautiful cyphering book prepared in North Carolina around 1780 by Martha and Elisabeth Ryan represented a worthwhile artistic achievement, it said little about the quality of the sisters' mathematics learning. We would contend however, that that would be a 21st-century perspective, and that in the 1780s the idea that a cyphering book should be as attractive as possible prevailed. Furthermore, since the arithmetic in the Ryan sisters' cyphering book went well beyond the elementary, the book suggested that, yes, females could cope with complex ideas in arithmetic. In addition, the book served as a reference manual for the rest of the authors' lives.

Augustus De Morgan (1853) contended that the recording of *abbaci* definitions, rules, cases, type problems and exercises in cyphering books produced too much mere copying and not enough genuine understanding. Almost certainly, Charles Dickens (1850) believed the same thing. Unfortunately, there are no hard data available to tell us how well students who prepared cyphering books actually learned mathematics. The closest we have to such data were those generated by the written examination in arithmetic that Horace Mann and his committee forced upon the Boston writing schools in the mid-1840s. However, we believe that the quality of the questions on that written examination was so poor that no reliable inferences can be made from the performance data about how well the students who attempted to answer the questions understood arithmetic. Intuitively, though, we believe that many students who learned mathematics through the cyphering/recitation approach succeeded in developing a sound problem-solving approach to the subject. Although we know of no *direct* evidence to support our viewpoint, the fact that the cyphering approach was used across the world for at least 600 years, suggests that for many learners it was an effective educational strategy.

Final Evaluative Comments

It could be mere coincidence that the move away from the cyphering approach seems to have accelerated in the 1840s, at the same time as Horace Mann introduced written tests to evaluate the effectiveness of students' learning and teachers' teaching. It could also be coincidence that at about that same time, normal schools, and graded elementary and high schools, were being introduced into American education. We believe that the weight of evidence suggests, however, that these events were connected with the demise of the cyphering-book tradition.

We are now sufficiently far removed in time from when the cyphering tradition prevailed to be able to recognize that some of the elements of that tradition were educationally defensible. It encouraged individualized learning, and individualized assessment. It fostered pride in students as they prepared "their books," the most beautiful cyphering books that they could possibly produce. It generated texts that, subsequently, students literally carried into their trade and professional lives for guidance. These advantages have been rarely taken into account by critics who have contended that after 600 years the tradition had become tired, and generated learning environments that were a source of boredom for many students (Allison, 1997; Richeson, 1935; Smith, 1973).

Any claim that cyphering books never served a worthwhile educational purpose, that they testified merely to how students mindlessly copied material from other existing documents or books—and because of that, held back students' learning and understanding of mathematics—needs to be scrutinized. As indicated above, it is likely that the cyphering approach *did* assist many students to understand key principles. Certainly, it gave many students reference books that they could consult when the time came to apply arithmetic in real-life situations. Knowing that they might have to refer to "their books" for guidance in the future usually meant that students did their best to make sure that what they wrote in their books was correct. Furthermore, the cyphering books concentrated on how to solve problems. Our analyses of the cyphering books that make up our PDS provided much evidence that students genuinely engaged in problem solving and spent little, if any, time on theoretical musings.

From that perspective, the corpus of cyphering books now lying in libraries and in collections across the nation should be seen as a national treasure. Like sewing samplers, often produced by schoolgirls, the cyphering books represent a rarity— education artifacts from a bygone age that were proudly prepared, celebrated, and treasured as representing sustained personal efforts to learn. Cyphering books, probably even more so than samplers, provided, for many of those who created them, trustworthy guides for important aspects of later living.

Warren Colburn (1830) spoke of a tendency for teachers who were using the cyphering approach to ensure that entries in their students' cyphering books were correct by actually solving the more difficult problems for the students. According to Colburn, this did not help the learner, for it produced nothing "except admiration of the master's skill in cyphering" (p. 30). From our own perspective, Colburn's comment provides a limited, even unfair, assessment of the learning that was likely to have occurred when a student prepared a cyphering book. Nevertheless, Colburn was speaking at a time when the cyphering approach was about to be tried in the balance and—at least from the perspective of its critics—found wanting.

Colburn was a Harvard graduate, based in Boston. He did not have much idea of what it was like to teach and to learn mathematics in the rural villages and on the farms across the United States of America. In most rural areas the only schools that existed were of the one-room, one-teacher variety (Goodrich, 1857). Teachers rarely had strong qualifications in mathematics, and learning aids like textbooks and blackboards were not readily available. Furthermore, in the New England states, and in New York and Pennsylvania, cyphering was usually done in winter months only. Given the need, throughout the 17th and 18th centuries, for families to maximize their labor force in order to survive and prosper in harsh economic conditions, cyphering was thought to be the only viable approach for mathematics education. It was not ideal, but it was useful because, after all, many teachers had kept the cyphering books that they themselves had prepared. Many students had access to cyphering books that a parent, or sibling, or some other member of the extended family, had prepared.

About a century ago, distinguished US scholars like Florian Cajori (1890), Louis Karpinski (1925), and David Eugene Smith (1925), when reflecting on the history of school mathematics in North America, paid much attention to the work and influence of famous textbook authors like John Bonnycastle, John Playfair, Thomas Dilworth, Nicolas Pike, Warren Colburn and Charles Davies. Cajori, Karpinski and Smith studied links between the mathematics curricula followed in schools and the mathematics entrance requirements and curricula of influential and prestigious colleges like Harvard, Yale, William and Mary, Columbia, Princeton, and West Point. They were not impressed with the heavy emphasis on copying and recitation associated with the cyphering tradition, and argued that the move to whole-class instruction and the development and use of better quality textbooks generated much improvement in teaching and learning. But, from our perspective, that approach to writing history proceeded from an unsatisfactory "present-to-the-past" form of historiography. We are not comfortable with the "great institutions," "great persons" and "great textbooks" approach to education history.

In this book, we have examined the development of mathematics education from the perspective of the students and teachers involved, taking account of the contexts in which school mathematics occurred. We have been particularly concerned to investigate what was possible, and what emerged, given the circumstances of place and time. This led us to examine the strong effects of the cyphering tradition on early school mathematics curricula and on instructional practices. We have been interested in the efforts of "ordinary" schoolmasters, like those of the long-forgotten Robert Lazenby, son of a Revolutionary War soldier from Maryland.

Lazenby was a teacher/farmer in North Carolina who, between 1813 and 1827, provided a form of mathematics education for children of plantation owners in his small subscription school (Lazenby, 1938). He died in 1828, at the early age of 42, but his 360-page cyphering book that he prepared between 1799 and 1802 lies at rest, among the Lazenby family papers in the Wilson Library at the University of North Carolina (Doar, 2006). It is well thumbed, and rather battered and scarred in appearance. Almost certainly, it was handled by Lazenby and by his subscription-school students on a daily basis. Lazenby's "book" would have provided the chief inspiration and sources of information for his students on a wide range of *abbacus*-like topics—from numeration to the various rules of three, and then fellowship, alligation, false position, and gauging. There were also sections dealing with elementary Euclidean geometry (definitions, and straight-edge and compass constructions), mensuration of superficies, and surveying.

We do not know how many "new" cyphering books were prepared by Lazenby's students, with Lazenby's book serving as the "parent." The idea of establishing lineage among early North American cyphering books is one which could usefully be taken up by future researchers.

A broader issue relates to the principal sources of mathematical knowledge for "the people" during the period 1607 and 1861. We would anticipate that dame schools, hornbooks, almanacs, cyphering books, apprenticeship schools, subscription and other local elementary schools and, above all else, day-to-day social interactions, were among the most important factors affecting what mathematics was learned, and how it was learned. Each of those factors was arguably more important than mathematics textbooks and mathematics professors in colleges.

It was the cyphering tradition which inculcated in those learning mathematics the idea that problems came in categories, and that each category could be subdivided so that particular rules and cases were appropriate. This realization led us to conceptualize what we called the IRCEE genre. The "promiscuous problem" sets found in many of the cyphering books in our PDS were intended to force students to ask themselves: "To which type of main problem category does this problem relate?" And, "which rule and case is the most appropriate?"

The cyphering tradition provided the curriculum glue that held together, yet constrained, the development of pre-college mathematics education in North America between 1607 and 1861. It was the force of that tradition which explains why the same arithmetic topics were studied, in more or less the same sequence, right across North America—in Boston, in New York, in Canada, and in small rural schools such as the one conducted, around 1820, by Robert Lazenby in North Carolina. The tradition provided teachers with a subconscious base for the intended, implemented and attained aspects of school mathematics, and in particular it provided the backdrop against which the PCA (problem-calculation-answer) genre, that still bedevils thinking and practice in school mathematics in the United States, arose.

Recommendations for Further Research

The cyphering approach has long since disappeared from the North American education scene, and so one might wonder whether there would be much point investigating issues associated with its demise. We believe, however, that because the demise came so abruptly—over about 20 years—and because a 600-year tradition that definitely affected what transpired in classrooms was done away with, the matter ought to be of considerable historical interest. Gaining an understanding of how, and why, this large-scale change to implemented curriculum happened so quickly might assist those seeking to energize school mathematics today.

It will not be easy for future researchers to investigate the effects of the cyphering tradition in North America unless brave measures are taken to consolidate and make more accessible to researchers the collections of cyphering books that already exist. Cyphering books have been given various names (cipher books, notebooks, student journals, copy books, arithmetic books, etc.), and are catalogued under different names in different libraries. Often, they are buried deep in collections of family papers, and a person interested in studying them must order them from the "front desk." Many of those employed to look "in the stacks" for requests from researchers do not know much about cyphering books, and therefore find it difficult to locate them.

We are not suggesting that cyphering books be taken out of the family history collections in which they are often to be found; rather, we believe that finding aids for cyphering book collections should be created in each library that holds a significant number of them. We are pleased to report that that is what happened after our work at the Phillips Library, within the Peabody Essex Museum at Salem, Massachusetts (see Gaydos & Kampas, 2010). The listing of cyphering books on the website of the Huguenot Historical Society (see Appendix B) is also a very helpful achievement. We expect that the summary of the manuscripts making up our principal data set provided in Appendix A in this book will also prove to be valuable. But much needs to be done. We are particularly concerned that numerous cyphering books have been lost in recent years, often because uninformed family members have decided to dispose of them.

Without further comment, we list five other important, and researchable, issues:

1. Why did the cyphering tradition linger longer in Great Britain than in North America, and how long did it survive in different European and non-European nations?
2. Were different cyphering traditions in evidence in the Northern and Southern states of the United States of America during the first half of the 19th century?
3. In the early 19th century, what proportion of cyphering books were prepared by school students, by college students, by apprentices, and by those being privately tutored?
4. To what extent did the greater availability of commercially-printed arithmetics in the 19th century contribute to the demise of the cyphering-book tradition?
5. To what extent was Van Egmond (1980) right when he suggested that the *abbaco* tradition lives on in modern school mathematics textbooks?

The completion of this book represents the end of the beginning of our research into the history of school mathematics. Our investigations into the cyphering era have caused us to reflect on school mathematics education in the 21st century. What stands out most for us relates to the concept of "ownership." When preparing cyphering books, many students came to feel that they "owned" what they were writing down. Not many students today feel that they own the mathematics that they are asked to learn. Can that be changed, and if so, how?

Author Biographies

Nerida Ellerton has been a Professor in the Mathematics Department at Illinois State University since 2002. Between 1997 and 2002 she was Professor and Dean of Education at the University of Southern Queensland and, between 1993 and 1997, Professor of Education at Edith Cowan University (1993–1997). Before that, she was Associate Professor at Deakin University, where, in 1991 and 1992, she was Director of the National Centre for Research and Development in Mathematics Education. Nerida holds two PhDs—the first is in Physical Chemistry (from the University of Adelaide), and the second in Mathematics Education (from Victoria University, Wellington). Nerida has written or edited 14 books. She has taught in schools, has served as an international consultant, and has had over 100 refereed scholarly papers published in the fields of mathematics education and chemistry. Between 1993 and 1997 she was editor of the *Mathematics Education Research Journal* and she is currently Associate Editor for the *Journal for Research in Mathematics Education*.

M.A. (Ken) Clements is a Professor in the Mathematics Department at Illinois State University. After teaching in schools, Ken completed his PhD at the University of Melbourne. He then taught in three Australian universities (Monash, Deakin, and Newcastle), and at Universiti Brunei Darussalam (1997–2004). In 1980 he was a visiting fellow at Cambridge University, and he has served as a consultant in India, Malaysia, PNG, South Africa, Thailand, the United Kingdom, and Vietnam. Ken has co-edited two international handbooks on mathematics education (1996 and 2003) and is editor for Springer's third international handbook. He has written or edited 24 books and more than 200 articles. With Nerida Ellerton, he co-authored a UNESCO book on mathematics education research. Ken is honorary life member of both the Mathematical Association of Victoria and the Mathematics Education Research Group of Australasia.

Appendix A: Summary of the Cyphering Books in the Principal Data Set

Appendix A provides brief descriptions of the 212 cyphering books in the principal data set (PDS). Each cyphering book is handwritten and was prepared in North America at some time during the period 1701–1861. The order of listing is chronological except in a few cases where manuscripts from the same family are presented one after the other. The abbreviation "m/s" is often used for "manuscript." This PDS constitutes the largest single collection of early North American cyphering books in the world. The Chicester/Pine (1701) and Prust (1702) cyphering books in the PDS are the two oldest US cyphering books that we have seen, or know of.

In Appendix A we provide a brief summary of each of the cyphering books in the PDS. Although, in most cases the information was gained from our analyses of the actual cyphering books, sometimes we have included information gleaned from Internet research. Unfortunately, there were some cases for which important information relating to a particular cyphering book (e.g., the name, or gender, of the writer, or when or where the manuscript was prepared) could not be determined. Such instances have been clearly indicated in the summary which follows. Occasionally, entries in this summary have been based on our best guesses, and we apologize if errors are found. We would appreciate it if readers who become aware of definite errors would notify us, bringing the errors to our attention.

Whenever a cyphering book was mentioned in the preceding text of this book, the number of the cyphering book—as indicated in the first column of the following summary—was given. In the column headed "Major Content & Level" we have indicated the main area of mathematics covered by individual cyphering books. Most of the manuscripts were concerned with arithmetic only.

By "Arithmetic 1" we mean arithmetic up to and including the single rule of three. The topics in this category included notation, numeration, the four operations of addition, subtraction, multiplication and division (collectively denoted by "4 opns"), compound operations (denoted by "cpd opns"), reduction, practice, percentage, simple and compound interest, vulgar fractions, decimals, and the single rule of three (denoted by "R of 3"). The category "Arithmetic 2" included the double rule of three, the rule of three inverse, proportion, discount, loss and gain, barter, tare and tret (sometimes denoted by "tare & t"), alligation, fellowship (denoted by "f'ship"), single and double position, gauging, annuities, equation of payments and involution and evolution (mainly square and cube roots). If a student had entries on the various rules of three then this has been indicated by "Rs of 3." The sub-category Arithmetic 2– did not include any of alligation, fellowship, and single and double position, involution and evolution, and the sub-category Arithmetic 2+ took account of all of those topics, including the various kinds of alligation, fellowship, and single and double false position. The category "Arithmetic 3" included at least two of permutations and combinations, arithmetic and geometric progressions (denoted APs and GPs), various forms of mensuration of 2-D and 3-D figures, and duodecimals.

Sometimes a cyphering book in the PDS was obviously not the first cyphering book prepared by the person who prepared it. The levels of penmanship and calligraphy that we have indicated (High, Medium, and Low, with sub-categories such as High– and Medium+) were based on our combined judgments.

Entries in the column headed "Linked to Math text" are based on those cases where we recognized that entries in cyphering books were obviously based on sentences, paragraphs, or examples in commercially-printed mathematics texts. We have a large private collection of 18th and 19th century mathematics textbooks (from the United Kingdom and from the United States of America and Canada), and, in some cases, it was possible for us to place alongside a cyphering book entry a "corresponding" section in a published textbook. Obviously, we will have missed instances of this, but we believe that throughout the cyphering era students did not usually copy material directly from published textbooks. Much more common was the practice of copying from an earlier cyphering book.

The acronyms IRCEE, PCA (often found in the "Comments" column of the following summary) refer to the *genres* of entries in the cyphering books (see, Ellerton & Clements, 2009b for definitions of these genres).

Summary of Cyphering Books in the Principal Data Set, North American Colonies/States

Period 1: Cyphering Books Commenced Before 1780

M/s ref #	Year/ Period	Name(s)	Gender M/F	Location	Number of Pages	Major Content, and Level	Range of Topics: to ...	Level of Penmanship	Dates on Pages?	Names on Pages?	Linked to Math Text?	Diagrams?	Comments
1	1701–1790s	Pine/Chicester families	M	New York, NY	32	Arithmetic 1 to 2	Cpd opns. to f'ship	Medium	Yes, occasionally	Yes (some)	No	No	There are sections from numerous members of the Pine/Chicester families. The first, clearly dated, section is, possibly, the oldest-known extant North American cyphering book.
2	1702–1703	Thomas PRUST	M	Not known	90	Arithmetic 2	R of 3	High	Year only	Yes (Some)	No	No	Very dark ink used, with very distinctive setting out.
3	1742–1743	Jonathan LIVERMORE Jr	M	Westboro, Worcester, NH	22	Arithmetic 1	Cpd opns to R of 3	Low	Yes	Yes	No	No	Includes an indenture document
4	c. 1767	Sally HALSEY	F	Morristown, NJ (?)	92	Arithmetic 2	Alligation	High	No	Yes (2 pages)	No	No	Early cyphering book by a female.
5	1770–1774	David TOWNSEND	M	Oyster Bay, NY	196	Arithmetic 2	R of 3, to Alligation	Medium	Yes	Yes (Often)	No	No	Red ink sometimes used.
6	1771	John GREY	M	Rhode Island, RI	110	Arithmetic 2	R of 3, to f'ship	Medium	Yes (Winter)	Yes (Some)	No	No	Perhaps, later a Revolutionary War colonel
7	1772	Silas MEAD	M	Greenwich, CT	78	Arithmetic 2	R of 3, to f'ship	Low	Yes (Winter)	No	No	No	Linked to m/s by Sarah and Silas Hervey Mead.

Period 1: Cyphering Books Commenced Before 1780 (Continued)

M/s ref #	Year/Period	Name(s)	Gender M/F	Location	Number of Pages	Major Content, and Level	Range of Topics: to …	Level of Penmanship	Dates on Pages?	Names on Pages?	Linked to Math Text?	Diagrams?	Comments
8	1774–1783	Amos LOCKWOOD	M	Rhode Island, RI	39	Arithmetic 1	4 opns	Medium-	No	Yes (Often)	No	No	Has Revolutionary War document, 1774, signed by William Ellery.
9	1775	Cornelius HOUGH-TALING	M	New Paltz, NY	250	Arithmetic 1 to 2+	Numeration, to double position	High	No	Yes	No	No	IRCEE, PCA, red underscoring. Well established Dutch family.
10	1776–1778	George WOOLLEY	M	Long Island, NY	132	Algebra	Cardano, APs, GPs,	High	No	Yes	No?	No	IRCEE, but not PCA. At "Richard Mallock's School"
11	1779–1784	Thomas BURLINGAME	M	Providence, RI	84	Arithmetic 1 to 2	R of 3, tare and tret	Medium	Yes (all year)	Yes (Some)	No	No	Very dark ink.

United States, Period 2: Cyphering Books Commenced During 1780–1789

M/s ref #	Year/Period	Name(s)	Gender M/F	Location	Number of Pages	Major Content, and Level	Range of Topics: to ...	Level of Penmanship	Dates on Pages?	Names on Pages?	Linked to Math Text?	Diagrams?	Comments
12	1780	Elkanah HALL	M	Mansfield, MA	22	Arithmetic 1	R of 3	Low	Yes (first)	Yes (first)	No	No	Mary A Hall wrote her name on numerous pages in Summer, 1839.
13	1780	William TURNBULL	M	Perth, NJ	91	Spherical geometry, trigonometry	Geometry, astronomy, trig, mechanics	High	Yes (first)	Yes (first)	No	Yes (v. good)	Later became a graduate of the Military Academy, and an engineer.
14	1784–1785	John ANDERSON	M	Virginia, VA	164	Arithmetic 1 to 2	Double R of 3	High	Yes (often)	Yes (often)	No	No	The teacher's name (James Pritchard) was inscribed on the front page
15	1785	Jonathan BUTLAR	M	PA (Bucks County?)	50 (marbled covers)	Arithmetic 1 to 2+	Double position	Medium	Yes (Winter)	Yes	No	No	This m/s was later taken over by Isaac Butler, and then Ann Butler.
16	1787	Isaac ATKINSON	M	Northumberland County, PA	8 (no cover)	Arithmetic 2	Tare and t, practice, f'ship	Low	Yes (Winter)	Yes (often)	No	No	Part of what was obviously a much larger m/s.
17	1788	Mahala GOVES	F	Hartford, CT	32	Arithmetic 1	R of 3	Medium	Once	Once	No	No	Many calculations with large numbers.
18	c. 1788	Seth Green TORREY	M	Cummington, MA	48	Arithmetic 1 to 2+	Double position	Low	No	Yes (often)	No	No	Seth Green Torrey was born April 17, 1773.

United States, Period 3: Cyphering Books Commenced During 1790–1799

M/s ref #	Year/Period	Name(s)	Gender M/F	Location	Number of Pages	Major Content, and Level	Range of Topics: to	Level of Penmanship	Dates on Pages?	Names on Pages?	Linked to Math Text?	Diagrams?	Comments
19	c. 1790	Unknown	Not known	Unknown	56	Arithmetic 2	Fellowship	Medium	No	No	Yes (Dilworth)	No	No mention of Federal currency
20	c. 1790	Unknown	Not known	Dighton, MA	29	Surveying	Division of land	Medium	No	No	No	Yes (Many)	Surveying was important in 18th C North America.
21	1792	Unknown	Not known	Unknown	32	Arithmetic 1 to 2	Currency conversion	Medium	No	No	No	No	Part of a larger m/s that is no longer present.
22	1792–1794	Olney BRAYTON	M	Foster, RI	90	Arithmetic 2	R of 3, to f'ship	Low-high	No	Yes (some)	No	No	Folder-like, soft cover, sewn. Poem on Rule of 3.
23	1785	Solomon TRUE	M	Unknown	60	Arithmetic 2	R of 3, to f'ship	Very low	No	Once	No	No	"13 months = 1 year" IRCEE, PCA
24	c. 1795	John KERLING	M	Germantown, NY	36	Arithmetic 1	4 opns, meas[1]	Low to medium	No	Yes (some)	No	No	Water-stained
25	1795–1798	HOUGH	Not known	Philadelphia, PA	36	Arithmetic 1 to 2	4 opns, measurement	Medium	Yes (Winter)	No	No	No	Cover is *Claypoole's American Daily Advertiser*, April 15, 1796. IRCEE, PCA, 2 color
26	1795–1798	Unknown	Not known	Northboro, MA	36 (8 still blank)	Arithmetic 2	Decimals, R of 3, f'ship	High	Yes (Winter)	No	No	No	
27	1795–1801	William SENTER	M	Portsmouth, NH	84	Arithmetic 1 to 2	4 opns, to f'ship, alligation	Mature medium/high	No	Yes	No	No	Probably began at age 10. Teacher's writing evident (Thomas Dearborn?)
28	1796	Benjamin TOLMAN	M	MA	90	Arithmetic 2, Navig[n]	R of 3, f'ship, sailing	Medium	No	No	No	Yes (3, for sailing)	Last 23 pages devoted to sailing. Includes log (Boston – West Indies)
29	1797	Lucy STARR	F	Middletown, CT	68	Arithmetic 1 to 2	R of 3, f'ship	High	No	Once	No	No	High calligraphy
30	1797, 1802	Hugh ROSS	M	Canada (?)	52	Arithmetic 2 to 3	Alligation, APs and GPs	Medium	No	Yes	No	No	M/s was purchased from Canada.
31	1798–1801	Levi JOHNSON	M	Marlborough, MA	80	Arithmetic 1 to 2	R of 3	Medium	Yes (some)	Yes (some)	No	No	Heavy emphasis on the practical.

United States, Period 4: Cyphering Books Commenced During 1800–1809

M/s ref #	Year/Period	Name(s)	Gender M/F	Location	Number of Pages	Major Content, and Level	Range of Topics: to . . .	Level of Penmanship	Dates on Pages?	Names on Pages?	Linked to Math Text?	Diagrams?	Comments
32	c. 1800	Thomas COOPER	M	Unknown	52	Arithmetic 2	R of 3, proportion	Medium	No	Yes	Yes (N. Pike)	No	This follows Pike's (1788) text.
33	c. 1800	Warner GREENE	M	Rhode Island	200	Arithmetic 1 to 3	APs and GPs	Medium-high	No	No	No	No	Although IRCEE, PCA evident, there is some emphasis on reasoning.
34	c. 1800	Warren Green	M	CT	60	Arithmetic 1	4 opns, meas¹	Medium	No	No	No	No	There is a section with Federal currency.
35	c. 1800	Unknown	Not known	Unknown	36	Arithmetic 2 to 3	Alligation alternate	High	No	No	Yes (N. Pike)	No	Examples are from Pike's (1788) text.
36	c. 1800	Unknown	Not known	New Haven, CT	144	Surveying	Geom, trig, astronomy, navigation	High	No	No	Yes (Gibson)	Many	Christ's Hospital influence?
37	c. 1795	Unknown	M (?)	Unknown	38	Geometry, trigonometry, surveying	Logs, trigonometry surveying	Medium high	No	No	No	Yes (Many)	Straight-edge/compass constructions, logarithms trig, surveying.
38	1800–1805	Unknown	Not Known	New York?	68	Arithmetic 1 to 3	To APs, GPs, double position	Medium+ to high–	No	No	Yes (Daboll)	No	There are 3 small quires, and handwritten extras
39	1800–1820	Mary TOLMAN	F	Weymouth, MA	66	Arithmetic 1	4 opns to simple interest	Medium+ to high–	No	No	No	No	Equal additions for subtraction. M/s is in good condition.
40	1801	Moses STEINER	M	Pennsylvania	150	Arithmetic 1 to 2	R of 3	High	No	At the front	No	No	German dialect used (Pennsylvania Dutch?)
41	c. 1801	Sarah PECKMAN	F	Dartmouth, MA	60	Arithmetic 1 to 2+	Business arithmetic	High	No	No	Daniel Adams	No	Used by later family members. No covers. Family tree at back.
42	1803–1804	John STEELE	M	Seneca County, NY	42	Arithmetic 1	4 opns, compound opns	Low	Yes. Winter	Yes	No	No	Rag soft cover and rag paper.

United States, Period 4: Cyphering Books Commenced During 1800–1809 (Continued)

M/s ref #	Year/Period	Name(s)	Gender M/F	Location	Number of Pages	Major Content, and Level	Range of Topics: to …	Level of Penmanship	Dates on Pages?	Names on Pages?	Linked to Math Text?	Diagrams?	Comments
43	c. 1805	Unknown author	Not Known	Vermont	62 (page torn, 2 sections)	Arithmetic 1 to 2-	Cpd opns, fractions, decimals	Medium	No	No	No	No	Used complement method for subtraction. Strong IRCEE and PCA. No covers.
44	c. 1805–1810	Joshua JAQUITH	M	Andover, Vermont	72	Arithmetic 1	R of 3	Low+	No.	No	No	No	Hand-sewn quires. Equal additions for subtraction.
45	1805–1810	Jacob RICHARDS	M	Weymouth, MA	236	Arithmetic 1 to 2	F'ship (S and D), etc	Medium	No	Yes	Yes (text not named)	No	Poor spelling. Became a physician. No covers. Complement method for subtraction.
46	1806	Sarah MEAD	F	Greenwich, CT	36 + 208 (2 m/s)	Arithmetic 1 to 2+	Rs of 3, S and D position, APs and GPs	Low+	Yes.	No	No	No	Linked to Silas (Sr) and Silas (Jr) m/s.
47	1807	Hiram EMERY	M	Eden, Maine	90 (rag paper)	Arithmetic 1 to 2-	4 opns, compound opns, to R of 3,	Low+	First page only	First page only	Yes, text not named	No	Dark calico hard covers. Content linked to fishing village environment.
48	1808	Adam RIGHTNAUER	M	Westphalia, Maryland	44 (original m/s was larger)	Arithmetic 1	4 opns, compound opns,	Medium	Yes. Winter months	No	No	No	German-background. Used complement algorithm for subtraction.
49	1808–1809	Henry KIMBALL	M	Hillsborough, NY	112	Arithmetic 1 to 2	Num^n, 4 opns, R of 3, practice	Medium	Yes. Winter	Yes	No	No	Sterling and Federal currencies assumed. No covers. Rag paper.
50	1808–1809	Stroud MCDOWEL	M	Weston, PA	70 +	Algebra	4 opns, word probs, APs and GPs	Medium to high	No	No	No	No	Attended Princeton, and became a judge.

United States, Period 4: Cyphering Books Commenced During 1800–1809 (Continued)

M/s ref #	Year/Period	Name(s)	Gender M/F	Location	Number of Pages	Major Content, and Level	Range of Topics: to …	Level of Penmanship	Dates on Pages?	Names on Pages?	Linked to Math Text?	Diagrams?	Comments
51	1808–1810	Barnet FLATTERY	M	Cattawissa, PA, then Lancaster, OH	72	Arithmetic 1	Reduction to inverse proportion	High	Yes, Winter	Yes, often	No	No	E. Barger, S'master in Cattawissa. Elijab Barger, S'master in Ohio
52	1809	Pierce Lacy TANNER	M	Maryland	332 (+ 128 blank)	Arithmetic 1 to 2+	Fractions, R of 3, f'ship, position	High, exquisite	No	No	Yes, Voster's arithmetic (Ireland)	No	Magnificent. Prepared by a father for a son (who would become a doctor). Full leather bound.
53	1809	Author unknown	Not known	Warwick, PA	96	Arithmetic 1	4 opns to R of 3	Medium to high	No	No	No	No	Equal additions for subtraction.
54	1809–1810	Josiah GARDNER	M	Hingham, MA	22	Arithmetic 1	4 opns to R of 3	Medium+	Winter	No	No	No	Rag paper with hard covers sewn around. IRCEE, PCA
55	1809–1813	Isaac HEATON	M	North Haven, CT	60	Arithmetic 1 to 2	4 opns to APs and GPs	High	Winter	Yes	Yes, Daboll	Yes (non-math)	Two colors used.

United States, Period 5: Cyphering Books Commenced During 1810–1819

M/s ref #	Year/Period	Name(s)	Gender M/F	Location	Number of Pages	Major Content, and Level	Range of Topics: to ...	Level of Penmanship	Dates on Pages?	Names on Pages?	Linked to Math Text?	Diagrams?	Comments
56	c. 1810	David CALDER-WOOD	M	Unknown	78	Arithmetic 2–3,	Interest, Rs of 3,	Medium	No	No	No	No	Soft, marbled covers, front cover detached.
57	c. 1810	Leonard LEVI	M	Bridgwater, MA, Harvard College	206	Arithmetic 2 to 3	Fractions, APs and GPs, mensuration	Medium	No	Yes, sometimes	Yes, Pike, Daboll, Walsh	Yes	Sections are sewn together, after the event.
58	c. 1810	Israel WILSON	M	New Jersey	144	Arithmetic 1 to 2	4 opns, reduction, f'ship,	Low to medium	Yes	Yes, often	No	No	2 soft-covered cyphering books.
59	c. 1810	Unknown author	Not known	Unknown	32	Arithmetic 2	Tare and t, alligation	Medium	No	No	No	No	Copying without understanding
60	c. 1810	Unknown author	Not known	Unknown	52	Arithmetic 2	Business topics (e.g., loss and gain)	Medium	No	No	No	No	Calculations without understanding
61	c. 1810	Author Unknown	Not known	Philadelphia, PA	77	Arithmetic 1 to 2	4 opns to R of 3, and f'ship	Medium	No	No	No	No	Pages numbered from p. 9 to p. 86.
62	c. 1810	Unknown author	Not known	New Hampshire	50	Arithmetic 1 to 2+	4 opns to APs and GPs	Medium+	No	No	Yes, Walsh	No	No covers, rag paper, not sewn
63	1810–1813	Silas Hervey MEAD	M	Greenwich, CT	224	Arithmetic 1 to 2	4 opns, Rs of 3, s and c interest, annuities	Low to medium	Yes. Winter	Yes, often	No	No	IRCEE, PCA. Witty comments made throughout. See other Mead manuscripts.
64	1810–1818	Jessamine FOSHAY	M	Mount Pleasant, NY	171	Arithmetic 1 to 2	4 opns to R of 3	Low+	Yes. Winter	Yes, often	No	No	Pages numbered. No covers.
65	1811–1812	Joel WOOSTER	M	Huntington, Connecticut	72	Arithmetic 1 to 2	4 opns, f'ship	Medium+	Yes. Winter	Yes	No	No	Prepared in Winter. Soft folder-like cover

United States, Period 5: Cyphering Books Commenced During 1810–1819 (Continued)

M/s ref #	Year/Period	Name(s)	Gender M/F	Location	Number of Pages	Major Content, and Level	Range of Topics: to ...	Level of Penmanship	Dates on Pages?	Names on Pages?	Linked to Math Text?	Diagrams?	Comments
66	1812	Silva POND	M	Franklin, MA	56	Arithmetic 1 to 2+	Rs of 3 (single and double)	Medium	No	No	No	No	IRCEE, PCA. Commercially-sold book (7.75" by 6.5")
67	1812	Unknown author	Not known	Maine	75	Arithmetic 2 to 3 geometry	Cube roots, perms, mensuration	Medium	No	No	No	Yes (in geometry)	Poems, newspaper clippings, etc., are pasted over 19 pages
68	1812–1813	William DARCY	M	Not Known	418	Arithmetic 2 then Practical	R of 3, practical arithmetic	Medium+	Yes Winter	Yes, occasionally	No	Yes	Very practical. Reference made to the use of a slide rule (by carpenters)
69	1812, 1840	Josiah, Sarah Ann LOVE	M F	Coventry, Rhode Island	250+	Arithmetic 1–3	4 opns to perms and combs.	Medium+	Sometimes	Yes	No	No	Made up of 5 sections (not sewn).
70	1813, 1816	Stites STEELE	M	Seneca County, New York	50	Arithmetic 1	Reduction, R of 3, inv proportion	Low-Medium	Yes often (Winter)	Yes, often	No	No	In worn, folder-like soft covers. See other Steele family books.
71	1813–1820	Robert Ross STEELE	M	Seneca County, New York	140	Arithmetic 2– to 2+	R of 3, position, f'ship	Medium	Yes, and rot only Winter	Yes, often	No	No	Worn, torn, soft cardboard cover. See other Steele family books.
72	1813–1818	George KINYON	M	Richmond, Rhode Island	205	Arithmetic 1–2+	Cpd opns to evolution, involution	Medium	Yes (Winter)	Yes	No	No	Linked with m/s by Daniel Kinyon (1824–1826)
73	1814	James KERSHAW	M	Massachusetts	54	Arithmetic 1	4 opns to R of 3	Medium	Yes, March	Yes, often	No	No	Water-staining on most pages
74	1814	John WILSON	M	Massachusetts	36	Arithmetic 1 to 2	Reduction to R of 3 inverse	High	Yes, January (all 36 pages)	Yes, often	No	No	No covers. PCA genre, but not many rules stated—mostly exercises.
75	1814–1827	Hiram FOUTS	M	Clark County, Indiana	300	Arithmetic 1–3	Arithmetic 1 to APs and GPs, etc.	Medium+ to High-	Yes, Winter	Yes	No	No	Used "complement" for subtraction. Rare early Indiana m/s.

United States, Period 5: Cyphering Books Commenced During 1810–1819 (Continued)

M/s ref #	Year/Period	Name(s)	Gender M/F	Location	Number of Pages	Major Content, and Level	Range of Topics: to …	Level of Penmanship	Dates on Pages?	Names on Pages?	Linked to Math Text?	Diagrams?	Comments
76	c. 1815	Jessie EASTMAN	M	Coventry, NH	74	Arithmetic 1 to 2	4 opns to Rs of 3	High-	No	No	No	No	Poor spelling. Subtraction by Complementary Algorithm IRCEE, PCA. Torn Rag cover, and rag paper. Authentic
77	c. 1815	John SHEPARDSON (1800–1878)	M	Franklin, MA	43	Arithmetic 1 to 2	Compound ops, to involution, evolution	Low to medium	No	No	No	No	
78	1815–1816	Thomas MOUNT	M	Maryland	422	Arithmetic 1 to 2	4 opns to permutations	High-	No	No	Yes, Jess's Assistant	No	Hard back, leather spine. IRCEE, PCA
79	1815–1816	Unknown author	Not known	New London, CT	30	Arithmetic 1 to 2	4 opns to R of 3, mensuration	High-	No	No	No	No	IRCEE, PCA Subtraction by the "complement" algorithm
80	1816	Richard WARNER M.A. Preceptor	M	Brandywine, Pennsylvania (Brandywine Boarding Sch)	208	Advanced Arith, Geometry, Surveying	Mensuration, trigonometry	High	Yes	Yes	No	Yes, many	Magnificent. Passed on through family. Teacher's cyphering book.
81	1816	Anrusa WIGGINS	F (?)	Portsmouth, New Hampshire	122	Arithmetic 1 to 2⁻	4 opns to Rs of 3	Low to medium	No	Yes, occasionally	No	No	Cover is Jan 27, 1816 issue of the *Portsmouth, New Hampshire Oracle*.
82	1817	Samuel Howard FAY	M	Boston, MA	160	Elemʸ Algebra	Elementary word problems	Medium	No (except on inside front cover)	Inside front cover	No	No	Picture of the Constitution on the inside of the front cover. Many pages have newspaper cuttings. Direct linked to Presidents Bush.
83	1817–1823	Thomas DOWNING	M	Lancaster, PA	50	Elemʸ Algebra, Geometry	2-D and 3-D geometry	High-	No	No	There is a reference to Dr Hutton	Yes, many beautiful diagrams	IRCEE, PCA. Perhaps this was a teacher's book. Hard cover, originally with marbled covers.
84	1818	Sarah EVANS	F	Chester County, PA	11	Arithmetic 1 to 2	Practice, R of 3	Medium	Yes, Summer	Yes, often	No	No	Probably prepared in local subscription school. Mother was German.

United States, Period 5: Cyphering Books Commenced During 1810–1819 (Continued)

M/s ref #	Year/Period	Name(s)	Gender M/F	Location	Number of Pages	Major Content, and Level	Range of Topics: to …	Level of Penmanship	Dates on Pages?	Names on Pages?	Linked to Math Text?	Diagrams?	Comments
85	1818	George PEIRCE	M	East Greenwich, Rhode Island	140	Arithmetic 2 to 3⁻	R of 3, f'ship, fractions, involution	High	No	No	No	No	Beautiful penmanship. Apparently, there were lots of Peirces in Greenwich, RI. IRCEE, PCA.
86	1818, 1836	Richard OSTRANDER	M	New Rochell, NY	64 + 21 = 85 Hard cover	Arithmetic 1 to 2	Reduction to f'ship	Medium to high	No	No	No	No	Last 21 pages, on Chemistry and Botany (probably c. 1836). Leather spine, marbled covers.
87	1818–1822	John S. HARNER	M	Philadelphia, PA	64	Arithmetic 1 to 2⁻	4 opns to R of 3	Medium	Yes Winter	Occasionally	No	No	Soft marbled covers are detached. Used by a number of people.
88	1818–1823	Thomas and Hannah CAVISS Samuel BROWN	M F M	Bristol, Maine	151	Arithmetic 2⁻ to 2⁺	Tare and t, involution, and evolution, f'ship	Medium	Yes mostly Winter but one Summer	Sometimes	No	No	Quite a few calculation errors. This m/s was developed by different people at different times.
89	1819	Pardon A. LAWTON	M	South Kingstown, Rhode Island		Arithmetic							In 1824, Pardon Lawton moved to Cross Creek, Washington County, Pennsylvania. Linked to Joseph Stoney Lawton (1841) m/s.
90	1819	I. SARIL	Not known	Unknown	84	Arithmetic 1 to 2	Reduction, Rs of 3, proportion	Medium	No	No	No	No	Sometimes scrappy, sometimes good. No covers.
91	1819–1823	Luther HAMPTON	M	Unknown	210	Arithmetic 1 to 2⁻	4 opns, R of 3	Medium	Yes Winter	Yes	No	No	Spelling is poor. IRCEE, PCA
92	1819–1829	Curtis, Clement, Bethiak SKOLFIELD, Eunice LINSCOTT	M M F F	Harpswell, Maine	146	Arithmetic 1 to 2⁻	4 opns, reduction, Rs of 3, involution, evolution, tare and t, f'ship, decimals	Low to medium	Yes Winter	Occasionally	Yes, Michael Walsh (4th ed.)	No	In Feb 1822, W. A. Lane was schoolmaster. Committee visited school, 30 students, including 4 Skolfied girls, were present. The "equal additions" algorithm was used for subtraction

United States, Period 6: Cyphering Books Commenced During 1820–1829

M/s ref #	Year/Period	Name(s)	Gender M/F	Location	Number of Pages	Major Content, and Level	Range of Topics: to …	Level of Penmanship	Dates on Pages?	Names on Pages?	Linked to Math Text?	Diagrams?	Comments
93	1820	Ann MATTISON	F	Solebury Seminary Philadelphia	36	Arithmetic 1 to 2–	4 opns, interest, practice	Medium+	No	No	No	No	Top one-thirds of 12 pages removed by previous owner (for art work)
94	c. 1820	Gersham, Arthur BIGELOW	M	New York (Robin Hill School)	119	Arithmetic 1 to 2+	4 opns to compound f'ship	Medium+ to High–	Yes	Yes	No	No	Strong IRCEE, PCA. Hard covers, leather spine. Used equal additions for Subtrn.
95	c. 1820	Abigail, COFFIN	F	Haverhill, MA	37	Arithmetic 1	Compound opns to R of 3	Medium	No	Yes	No	No	IRCEE, and especially PCA. Card, folder-like cover.
96	c. 1820	Matilda CUMING (?)	F	Providence RI (?)	28	Arithmetic 1	4 opns, reduction	Low to medium	No	No	Yes, Daboll	No	IRCEE, PCA. Equal addition for subtraction.
97	c. 1820	Stephen B. CURRIER	M	East Kingston, NH	56	Arithmetic 2 to 2+	Alligation, permutations	Medium –	No	Yes	No	No	Old newspaper covers
98	c. 1820	John WALTERS	M	Unknown	7	Arithmetic 1	Compound operations	Medium	No	No	No	No	IRCEE, PCA
99	c. 1820	Unknown author	Not known	Rhode Island	30 (2 m/s)	Arithmetic 1	4 opns, compound opns	Medium	No	No	Yes, Daboll	No	Equal addition for subtraction (but complement for compound opns!)
100	1820–1821	Abigail B. MOUNTS	F	Unknown	44	Arithmetic 1 to 2	4 opns to f'ship	Medium+	Yes, all the year	No	No	No	Soft card covers. Mostly exercises. PCA
101	1820–1828	William BUNN	M	Orange County, NY	55	Arithmetic 1 to 2–	4 opns to R of 3	Low	Yes, Winter	Yes	No	No	Newspaper covers.
102	1820–1834	Joseph W. THOMPSON, Rhodiland JOHNSON 2 Unknown authors	M F	Maine	100 (in 4 m/s)	Arithmetic 1 to 2	4 opns to R of 3, alligation, position.	Medium–	No	No	No	No	One used equal additions for subtraction, another the complement method. Rag paper. Only one m/s had a cover (which was rag paper)
103	1820s	Michael SIEGRIST	M	Silver Spring, PA	110	Arithmetic 2 to 2+	R of 3, double position	Low to medium+	No	Yes, often	No	No	Three separate sections, no covers, rag paper,

United States, Period 6: Cyphering Books Commenced During 1820–1829 (Continued)

M/s ref #	Year/Period	Name(s)	Gender M/F	Location	Number of Pages	Major Content, and Level	Range of Topics: to ...	Level of Penmanship	Dates on Pages?	Names on Pages?	Linked to Math Text?	Diagrams?	Comments
104	1821–1822	Addison METCALF	M	Lisbon, Maine	17	Arithmetic 1 to 2+	Compound opns to progressions	Medium + to high –	No	Occasionally	Yes. N. Pike	No	Questions are often taken from Pike's arithmetic. Elaborate calligraphy.
105	1821–1822	Jacob WENTWORTH	M	Milton, MA	72 (in 2 sections, numbered pages)	Arithmetic 2 to 2+	S and C interest, double posn, mensuration	Low + to medium–	Yes, occasionally (Winter)	Yes, occasionally	No	No	No covers. Jacob was aged 18 to 19 at the time.
106	1821–1823	Elizabeth HOLBROOK	F	Sherburn(e), MA	66 + poem ("Piety")	Arithmetic 1 to 2	4 opns to double f'ship	High, stunning	Yes, occasionally	No	No	No	Magnificent calligraphy, penmanship. Soft grey covers. Pure IRCEE, PCA.
107	1821–1827	Mary BURLINGAME	F	Peters Burg, RI	166	Arithmetic 1 to 2+	Compound opns. to alligation, APs and GPs	Medium to high–	Yes, all year	Yes, occasionally	Yes. Daboll	No	Wide range of topics. See other Burlingame manuscript summaries.
108	1822	James C. HUNTINGTON	M	Neville, OH, then Bracken Co., Kentucky	8 (– other pages)	Arithmetic 1 to 2+	Compound operations	Low	Yes, Winter	Yes, occasionally	No	No	Numerous newspaper cuttings. Scrappy.
109	1822–1827	Albert HAWKINS Augustine HAWKINS	M F	Unknown	450+	Arithmetic 1 to 2+	4 opns, to gauging, position, APs and GPs	Medium	Yes, all the year	Often	No	No	Loose leather covers. Sections written by both Albert and Augustine.
110	1823	William H. WELLS	M	Columbus, Ohio	70	Arithmetic 1 to 2+	4 opns, to alligation, position, APs and GPs	Medium+	Yes, Winter	Occasionally	Yes Daboll	No	IRCEE, PCA

United States, Period 6: Cyphering Books Commenced During 1820–1829 (Continued)

M/s ref #	Year/Period	Name(s)	Gender M/F	Location	Number of Pages	Major Content, and Level	Range of Topics: to ...	Level of Penmanship	Dates on Pages?	Names on Pages?	Linked to Math Text?	Diagrams?	Comments
111	1823–1827	Ursula WILSON	F	Pennsylvania	32	Arithmetic 1	Compound operations	Medium	No	No	No	No	No cover. Ursula may have been blind.
112	1824	Anderson NEWMAN	M	Shenandoah, Virginia	40	Arithmetic 2+ to 3–	GPs, duodecimals, annuities	Medium	No	No	No	No	This comprises pp. 341–381 of some larger m/s. Newman would become a politician. IRCEE, PCA.
113	1824–1826	Bertha JETTY	F	Boston, MA	42	Arithmetic 1	4 opns, compound opns	Low+	No	No	No	No	IRCEE, PCA. Advert for early commercially-printed cyphering books
114	1824–1826	Daniel KINYON	M	Richmond, Rhode Island	132	Arithmetic 1–2	Cpd opns to Equation of payments	Medium	Yes, Winter	Yes, often	No	No	Daniel often included poems, etc. His spelling was poor. M/s is linked to George Kinyon's (1813–1818) m/s.
115	1824–1830	William TOWNSEND	M	Oyster Bay, New York	212	Arithmetic 1 to 2+	Compound opns, to d. position	Medium+	No	No	No	No	IRCEE, PCA. Uses complement algorithm for subtraction.
116	1825	Abraham POST	M	Massachusetts	110	Arithmetic 2	Reduction, duodecimals, annuities	Medium+	Yes Winter	No	No	No	Very worn, marbled soft covers.
117	1825 (also 1790)	Daniel H. SANFORD, Nathan C. SANFORD	M M	New York	56	Arithmetic 1 to 2+	R of 3, practice, interest f'ship	Medium	Yes, Winter	Yes	Yes, Daboll	No	Nathan C Sanford , Daniel's father, was a VP nominee for the 1824 election. IRCEE, PCA.
118	c. 1825	Joel FOSTER	M	Meriden, CT	32	Arithmetic 1 to 2–	Barter, interest, discount	Medium –	No	Sometimes	No	No	Rag paper covers. Foster would become a judge.
119	1825–1826	Joseph SEIBERT	M	Maryland	178	Geometry, Surveying, Trig	Much pure and applied.	Medium+ to high–	Yes, Winter	Yes, often	No	Yes (high quality)	Straight-edge compass constructions. Excellent diagrams. Hard covers

United States, Period 6: Cyphering Books Commenced During 1820–1829 (Continued)

M/s ref #	Year/Period	Name(s)	Gender M/F	Location	Number of Pages	Major Content, and Level	Range of Topics: to …	Level of Penmanship	Dates on Pages?	Names on Pages?	Linked to Math Text?	Diagrams?	Comments
120	1825–1828	Elijah Allen ROCKEFELLER	M	Hunterdon County, NJ	54	Arithmetic 1 to 2–	4 opns to Rs of 3	Low to medium-	Yes, Winter	Yes	No	No	Poor spelling throughout.
121	1825–1829	Samuel ANDERS	M	Philadelphia, PA	88	Arithmetic 1 to 2	Compound opns to f'ship	Medium to high-	Yes, Winter	Yes, often	No	No	Soft marbled covers, good condition. IRCEE, PCA.
122	1825–1829 and 1846	Peter, Michael, Jacob, Gideon, David SINK	5 Males	Davidson County, North Carolina	32	Arithmetic 1	4 opns, reduction, compound opns.	Low	Yes, Winter	Yes	No	No	Low standard of penmanship, calligraphy throughout
123	1826–1829	John TINKHAM	M	Fairhaven, MA	186	Arithmetic 1 to 2+	Compound opns to APs, GPs, perms and combs	Medium+	No	Yes	Yes, Adams, Daboll, Walsh	No	Includes an excellent letter about the examination committee.
124	1827 and 1837	Luther EAMES	M	Bridgewater, MA	84	Arithmetic 1 to 2–	4 opns, reduction. R of 3	Low	No	No	No	No	Rag paper, used equal additions for subtraction.
125	1827	Levi JONES	M	Marlboro, NH	72	Arithmetic 1 to 2	Numeration to double R of 3.	Medium	No	No	No	No	Soft marbled covers. M/s is in good condition.
126	1827–1828	John GEIGH	M	Pennsylvania	28	Arithmetic 1	4 opns to R of 3	low	No	Yes, 26 times	No	No	Hard covers. John became an Army officer.

United States, Period 6: Cyphering Books Commenced During 1820–1829 (Continued)

M/s ref #	Year/Period	Name(s)	Gender M/F	Location	Number of Pages	Major Content, and Level	Range of Topics: to ...	Level of Penmanship	Dates on Pages?	Names on Pages?	Linked to Math Text?	Diagrams?	Comments
127	1827–1831	David Sharp AUSTIN	M	Clinton County, Ohio	222	Arithmetic 1 to 2	4 opns to Rs of 3, alligation	Medium	No	Yes	No	No	Poor condition, rag paper quires, stitched together. Partly burnt.
128	1828	James EMERY	M	Not Known	102	Arithmetic 1 to 2+	4 opns, to alligation, f'ship	Medium	No	No	No	No	Many of the leaves are loose. Hard cover with name in gilt on outside.
129	1828	John Wesley PIERSON	M	Talbot County, Maryland	230 (pages numbered)	Mensuration	Conics, trig, logarithms.	High	Yes. March 31 to Sept 19	No	Yes, Bonnycastle	Yes	Teacher was John A. Getty. 230 pages completed in 6 months. Summer.
130	1828	Netty Ann STEELE	F	Seneca, New York	18	Arithmetic 1	Compound opns, reduction	Low to medium	Yes Winter	Occasionally	No	No	Very elementary, mostly PCA. Emphasis on exercises. Part of Steele family set of m/s.
131	1828–1829	Sarah EAMES	F	Bridgewater, MA	18	Arithmetic 1 to 2–	4 opns to R of 3	Low+	No	No	No	No	See, also, the Luther Eames m/s.
132	1829–1832	William KING	M	Massachusetts	132 (in 2 manuscripts)	Arithmetic 1 to 2+	4 opns, compound opns	Medium	Yes, Winter	Often	No	No	Elementary arithmetic only. Linked with Philip and Henry King manuscripts.
133	1829–1835	Samuel ARNOLD	M	Rhode Island	156	Arithmetic 1 to 2+	4 opns to f'ship, then fractions.	Medium	Yes, Winter	Occasionally	No	No	Two folder-like card covers. Used equal additions for subtraction

United States, Period 7: Cyphering Books Commenced During 1830–1839

M/s ref #	Year/Period	Name(s)	Gender M/F	Location	Number of Pages	Major Content, and Level	Range of Topics: to ...	Level of Penmanship	Dates on Pages?	Names on Pages?	Linked to Math Text?	Diagrams?	Comments
134	1830	Evan BROWN	M	Canada	120	Arithmetic 1+ to 2+	Decimals	High–	No	No	No	No	Hard marbled covers. Surds a topic (but not in the US)
135	c. 1830	Albert CHAPMAN	M	Westerly, RI	16	Arithmetic 1	4 opns, compound interest	Low	No	No	No	No	Scrappy, includes indenture document.
136	c. 1830	John G. REID	M	Nova Scotia, Canada	52	Arithmetic 2 to 2+	alligation, f'ship	Medium+ to high–	No	No	No	No	Horn-made vellum soft covers. Perhaps a teacher's cyphering book.
137	c. 1830	SOLISTRE (?)	F (?)	Rochester, NH	82	Arithmetic 1+ to 2+	Vulgar fractions to APs and GPs	Medium+ to high–	No	No	No	No	Torn soft, rag covers. Rag paper
138	c. 1830	Hannah P. WINKLEY	F	Gilmanton, NH	20	Arithmetic 1 to 2,	Fractions, Rs of 3	Medium	No	No	No	No	Scrappy. IRCEE and PCA. No covers.
139	c. 1830	Unknown author	Not known	Unknown	12	Arithmetic 2	Double position	Medium	No	No	Yes, Thomas Smiley	No	All problems were from Thomas Smiley's *Federal Calculator*: PCA.
140	1830–1831	Albert PLUMER	M	Newbury, MA	268	Arithmetic 1 to 2+	Compound opns, to perms and combs	Medium + to high–	No	No	No	No	Soft rag covers. Topics often repeated. The first of two large Albert Plumer cyphering books.
141	1830–1833	Nathan CLIFTON	M	Ohio	335	Arithmetic 1 to 2+	Compound opns to double position	Low + to medium–	No	No	No	No	Some nice colored drawings (not math). Soft rag covers. Later became Confederate major in Civil War
142	1830–1835	David S. BROCK	M	Thompson, CT	76	Arithmetic 1+ to 2	Reduction, Rs of 3, f'ship	Medium+ to high–	No	No	No	No	Soft, commercial cover (sold by Caleb Burbank, Millbury, MA.
143	1830–1835	Timothy GOOD	M	New York	56	Arithmetic 1 to 2–	4 opns to R of 3	Medium+ to high–	No	No	No	No	Equal additions for subtraction. Hard cover. Vellum spine cover.

United States, Period 7: Cyphering Books Commenced During 1830–1839 (Continued)

M/s ref #	Year/Period	Name(s)	Gender M/F	Location	Number of Pages	Major Content, and Level	Range of Topics: to ...	Level of Penmanship	Dates on Pages?	Names on Pages?	Linked to Math Text?	Diagrams?	Comments
144	c. 1831	Mary TALCOTT	F	Hartford Female Academy, CT	48 numbered	Arithmetic 1	4 opns, to compound opns	Medium+ to high–	No	No	No	No	Soft card covers. Link with Catherine Beecher Stowe
145	1831–1832	John Hillary NEWMAN	M	Alabama	24	Arithmetic 1 to 2–	Compound opns, to R of 3	Medium–	Yes, Winter	No	No	No	Nice calligraphic headings. IRCEE
146	1831–1833	Stephen C. BROWNING	M	South Kingston, RI	125	Arithmetic 1 to 2	Compound operations, f'ship	Medium+	Some. All year around.	No	Yes, Daboll	No	Joseph R. Knowles, writing master. Pages numbered.
147	1831–1833	Albert PLUMER	M	Newbury, MA, then, perhaps, completed in Ohio	333	Surveying	Copied from a surveying text	High–	No	No	Probably. Gibson.	No	Soft, newspaper covers. It seems that Albert had a tutor to help him. The second of 2 large m/s from Albert.
148	1831–1834	Peter S. SLOCKBOWER	M	Essex County, NJ	185 (pages numbered)	Arithmetic 1 to 2+	Compound opns to f'ship	Medium+ to high–	Yes, Winter	No	Yes. Daboll	No	Peter died in 1864, aged 46. He had 10 children. Had a tutor
149	c. 1832	Betsey COOK (Hathaway)	F	Saline, Michigan (moved there, from NY, in 1832)	14	Arithmetic 1	4 opns, Federal money	Medium+	No	No	No	No	IRCEE. PCA. Card covers.
150	1832	Moses WILDES (aged 22)	M	Ipswich, MA	39	Algebra	Equations, no graphs	High–	No	No	No	No	This soft-covered book was probably the exercise book that accompanied the bigger cyphering book.
151	1833	John MAJOR	M	Unknown	38	Logarithms and Trigonometry	Logarithms, oblique trig, surveying	High–	Yes, all year around	Occasionally	Yes, Jess's *Surveying*	Yes (water-colored)	Unusual emphasis on logarithms, trig, with beautiful colored diagrams.

United States, Period 7: Cyphering Books Commenced During 1830–1839 (Continued)

M/s ref #	Year/Period	Name(s)	Gender M/F	Location	Number of Pages	Major Content, and Level	Range of Topics: to ...	Level of Penmanship	Dates on Pages?	Names on Pages?	Linked to Math Text?	Diagrams?	Comments
152	1833	Ambrose M. WOODWARD	M	Taunton, MA	36	Arithmetic 1 to 2	Reduction to double f'ship	Medium	No	No	No	No	Ambrose was 17 years of age. IRCEE, PCA.
153	c. 1833	Benjamin Franklin CHAMPION	M	West Springfield, MA	66	Arithmetic 1 to 2–	4 opns to Rs of 3	Medium	No	No	No	No	Light ink. Folder-like card covers. IRCEE and PCA. Equal Additions for Subtn.
154	1833–1834	Philip KING	M	Boston, MA	32	Arithmetic 1	4 opns to R of 3	Medium	Yes, Winter	Yes	No	No	Linked to the King family cyphering books. There are many spelling errors.
155	1833–1834	Reuben MATTISON	M	Bolton, Warren County, NY	50	Arithmetic 2	Loss and gain, barter, Rs of 3	Low	Yes, Winter	No	No	No	PCA. No covers. This was originally part of a larger manuscript.
156	1833–1837	Joseph HENRY Susanna LESHER	M F	NY (?)	94	Arithmetic 1 to 2	Compound opns, barter f'ship	Medium	No	No	No	No	Susanna Lesher may have inherited Joseph Henry's cyphering book
157	1833–1837	Elias MOUNT	M	Frederick County, Maryland	76	Arithmetic 1 to 2–	4 opns to R of 3	Low+	Occasionally Winter	No	Yes, Jess's *Assistant*	No	Another Mount family m/s. IRCEE, PCA. They had all used Jess's *Assistant*.
158	1834	Elizabeth MOUNT	F	Philadelphia, PA	30	Arithmetic 2–	R of 3, tare and t, interest	Medium	No	No	No	No	In a soft-covered, commercially printed cyphering book.
159	1834–1835	Samuel DENTON	M	Brier Island, Nova Scotia, Canada	160	Arithmetic 2– to 2+	Barter, interest, APs and GPs	Medium	Often, Winter	Yes, often	No	Yes, a few	Denton was a descendant of loyalists from New York. Poor condition.
160	1835	Albion Tipton BLAIR (Aged 24)	M	Columbus, McMinn County, Tennessee	134	Arithmetic 2	Rs of 3, f'ship, S and D position	Low, mature (a business man)	Yes, June, July	No	No	No	Teacher's name was John Carter (receipt to him for $4 tuition). No covers. Very dark ink.

United States, Period 7: Cyphering Books Commenced During 1830–1839 (Continued)

M/s ref #	Year/Period	Name(s)	Gender M/F	Location	Number of Pages	Major Content and Level	Range of Topics: to …	Level of Penmanship	Dates on Pages?	Names on Pages?	Linked to Math Text?	Diagrams?	Comments
161	1835–1837	Sarah S. MATTISON	F	Philadelphia, PA	12	Arithmetic 2–	Single, Double R of 3	Low+	No	No	No	No	Blue lined pages within soft commercially-printed covers. Also, blue ink. IRCEE, PCA. No covers.
162	c. 1835	Unknown author	Not Known	Athol, MA	32	Arithmetic 1 to 2	Compound opns, to d f'ship	Medium	No	No	No	No	
163	1836–1837	Mary Ann BAKER	F	Maryland	6	Arithmetic 1	Compound operations	Low+	No	No	No	No	On page 6 there are some recreational arithmetic problems, including a 9 by 9 magic square.
164	1836–1837	John M. FILLER	M	Boonville, Missouri	123 (pages were numbered)	Arithmetic 2	Barter, interest, f'ship	Low to medium–	Yes, Winter	Yes, often	Yes, Pike and Jess	No	There is an index to Pike and to Jess at the back of the first m/s.
165	1836–1843	William BRUNNER	M	Petersburg, NY	202	Arithmetic 1 to 2	4 opns to Rs of 3	Medium	Yes, Winter	Yes, often	No	No	Over the years, many topics were repeated. Equal additions for Subtraction.
166	1837	William HALE	M	Newbury, MA	78	Trigonometry, Navigation	Euclid, trig, navigation, journal log	High. magnificent	No	No	No	Yes, some colored	Superb penmanship, calligraphy, diagrams. Christ's Hospital-like.
167	1838–1838	Linley BROWN	F	Philadelphia, PA	136 (12.25″ by 8.5″)	Arithmetic 2 to 3 –	Vulgar fractions to permutations	Medium+	Yes (Winter)	Yes	No	No	Soft grey cover. IRCEE and PCA genres.
168	1837–1838	Levi JOHN	M	West Vincent, PA	124 (in 3 m/s)	Mensuration	Conics, builder's arithmetic	High–	Yes, not only Winter	No	No	Yes, some colored	Jesse E. Philipe, Preceptor
169	1837–1844	John W. ARMEL	M	Unity Township, PA	210	Arithmetic 2 to 2+, Geometry	R of 3, to perms and combs.	Medium–	No	Occasionally	No	A few.	Teacher was Samuel Armor, AM. No covers. Some diagrams.

United States, Period 7: Cyphering Books Commenced During 1830–1839 (Continued)

M/s ref #	Year/Period	Name(s)	Gender M/F	Location	Number of Pages	Major Content, and Level	Range of Topics: to …	Level of Penmanship	Dates on Pages?	Names on Pages?	Linked to Math Text?	Diagrams?	Comments
170	1838	George KRESHLER	M	Bucks County, PA	44	Arithmetic 2	Tare and t, to f'ship	Medium+	Yes, Feb. 1838	No	No	No	44 pages completed, neatly, in one month. IRCEE, PCA.
171	1838–1842	John Harry AKERS	M	Bush Creek Valley, PA	150	Arithmetic 1 to 2+	4 opns to alligation, f'ship	Medium	Yes, April 18 (Winter?)	Yes, often	Yes (Stephen Pike)	Yes, (but math irrelevant)	Mr Richard Allinder was tutor. Soft, worn covers sewn around rag paper quires.
172	1839	Jarvis H. BARTLETT	M	New Jersey	72	Arithmetic 2– to 2	Interest barter, loss and gain	Low+ to medium–	No	Occasionally	No	No	Would become a Republican politician. Soft covers. Hand-stitched, rag-paper quires.
173	1839	George A. HOWLAND	M	Fall River, MA	20	Arithmetic 1 and Ely Algebra	Reduction, and algebra equations	Medium+	No	No	Yes, Greenleaf	No	The writing is in pencil. Book was commercially printed in Fall River, MA.
174	1839–1840	Ann BEARDSLEY	F	New Paltz, NY	239	Arithmetic 1 to 2+	4 opns, to f'ship, alligation, APs and GPs	Medium+	Yes. Entries completed in 1 Winter!	No.	No	No	Location but known, but Beardsley was a common name in New Paltz. No covers. Sections sewn together.
175	1839–1840	ZIMMERMAN	Not known	Pennsylvania (German-language m/s)	352	Arithmetic 1+ to 2+	Fractions, R of 3,	High, exquisite	No	No	No	No	There are 6 other, smaller, German-language m/s accompanying this one.
176	1839–1840	Oliver P. SIMPSON	M	Stark County, Ohio	152	Arithmetic 1 to 2	4 opns to tare and t, R of 3	Medium+	Yes, for one whole Winter	Yes, often	No (Pages are numbered)	No	Calico cover, sewn-in quires, much content covered for one Winter.
177	1839–1841	Adam KISER	M	Latrobe, PA, and Baltimore, MD	94	Arithmetic 2	R of 3, to f'ship and APs and GPs	Medium	Yes, Winter only	Yes often	No (pages are numbered)	No	Within the marbled hard covers (which are detached) are incredible, mottled soft covers. IRCEE, and PCA.
178	1839–1842	Henry KING	M	Philadelphia, PA	62	Arithmetic 1	4 opns to R of 3	Medium	Yes, Winter	Yes	No	Yes	Linked to the King family cyphering books.

United States, Period 8: Cyphering Books Commenced During 1840–1849

M/s ref #	Year/Period	Name(s)	Gender M/F	Location	Number of Pages	Major Content, and Level	Range of Topics: to …	Level of Penmanship	Dates on Pages?	Names on Pages?	Linked to Math Text?	Diagrams?	Comments
179	1840	Lewis FREE	M	Unknown	20	Arithmetic 1 to 2–	Compound opns, tare and tret	Medium+	No	Yes, often	No	No	No covers. Originally part of a larger manuscript.
180	1840	Lemuel SPEAR (12 years old)	M	Quincy Centre School, Haverhill, MA	140 (2 hard-bound m/s)	Arithmetic 1 to 2+	4 opns to double f'ship	High (for 12-year-old)	Yes, Winter	Yes, often	No	Yes (for mensuration)	Equal additions, for subtraction. Teacher is Parker Johnson. "My father did this when he was a boy."
181	1840–1842	David HESS	M	Baltimore, MD	60	Arithmetic 1 to 2	4 opns to APs and GPs	low	No except for the year)	Yes, often	No	No	No covers, in very poor condition. Some parts of pages are eaten away.
182	1841	Joseph Stoney LAWTON	M	Pennsylvania	84 (13.5" by 8")	Arithmetic 1 to 2	Cpd opns to involution, evolution	Medium+	Yes	Yes	No	No	IRCEE and PCA genres. Linked to Pardon Lawton (1819) m/s.
183	1841–1844	Theodore Hearmon LESHER	M	New York	74 (in three m/s)	Arithmetic 1 to 2	Reduction to R of 3, barter, loss and gain, f'ship	Low	Yes, Winter only	Yes, often	No	No	Part of the Lesher family manuscripts. "Wrote by Owen Walter in 1844" is inscribed on the front of smallest m/s.
184	1841–1846	John NIXON	M	Pennsylvania	172	Arithmetic 2– to 2+	R of 3, to APs and GPs	Medium	Yes, Winter only	Occasionally	No	No	Colored headings. Commended by school committee.
185	1842	Daniel VALENTINE Samuel WEEKS	M M	Schenectady, NY	54	Arithmetic 2– and algebra	R of 3, then elementary algebra	Low	Yes, Winter only	Occasionally	Yes, Willett's and Adams' *Arithmetic*	No	Weeks could be a Revolutionary War veteran. Sewn, full leather cover.
186	1843–1844	Robert D. HALL	M	Portsmouth, NH	32	Geometry, trigonometry, surveying	Euclidean constructions, trig, surveying	Medium+	No	No	No	Yes	The construction arcs are very clear. This has folder-like covers.

United States, Period 8: Cyphering Books Commenced During 1840–1849 (Continued)

M/s ref #	Year/Period	Name(s)	Gender M/F	Location	Number of Pages	Major Content, and Level	Range of Topics: to ...	Level of Penmanship	Dates on Pages?	Names on Pages?	Linked to Math Text?	Diagrams?	Comments
187	1843–1847	Gertrude DEYO	F	New Paltz, NY	125 (Pages are numbered)	Arithmetic 1 to 2	4 opns, compound opns, Rs of 3, f'ship	Medium	No	No	No	No	Complementary algorithm for subtraction. Blue ink, blue-lined pages. Some pages are moth-eaten.
188	1844–1845	William Cook PEASE (25, in 1844)	M	Martha's Vineyard, MA	60	Geometry, navigation (keeping a log)	Navigation notes for Schooner Van Buren	Medium (mature)	Yes, as part of the log	Yes, regularly.	No	No	Pease would become the first director of the US Coast Guard. This is full of calculations made at sea.
189	1844–1847	Sarah E. CHAPIN	F	Enfield, MA (Aged 10 in 1844)	124	Arithmetic 1 to 2	4 opns to f'ship and evolution	Medium+	Yes, Winter.	Yes, occasionally	No	No	Used equal additions for subtraction. Hard green covers with gilt inscription. Impressive for young girl.
190	1846	Samuel P. BARTOLET	M	Pennsylvania	90 (Was this the exercise book?)	Arithmetic 2	Rs of 3, equation of payments, f'ship	Low+	Yes, one Winter.	Often	No	Some	No covers. The m/s was completed in January and February, 1846. PCA, solutions to problems.
191	1846	Franklin LAMB	M	North Carolina?	124	Arithmetic 1 to 2+	Reduction to perms and combs	Medium+	No	No	No	No	Soft rag covers, blue-lined paper, blue ink.
192	1846–1847	Emoline HAMPSHIRE	F	Baltimore, MD	60	Arithmetic 1 to 2–	Fractions, Rs of 3, barter	Low+ to medium–	Yes, Winter	Yes, often	No	No	Hard covers, leather spine, blue ink, lined pages. PCA genre.

United States, Period 8: Cyphering Books Commenced During 1840–1849 (Continued)

M/s ref #	Year/Period	Name(s)	Gender M/F	Location	Number of Pages	Major Content, and Level	Range of Topics: to …	Level of Penmanship	Dates on Pages?	Names on Pages?	Linked to Math Text?	Diagrams?	Comments
193	1847	Betsy T. KING	F	Virginia	13	Geometry, surveying	Euclidean constructions, surveying	Medium+	Once	Once	No	Yes, many	This is a rare geometry m/s prepared by a female.
194	c. 1848	Unknown teacher	Not known	Philadelphia, PA	70	Trigonometry, navigation	Solution of triangles, different types of sailing	Medium–	No	No	Davies' *Surveying*, Gummere's *Surveying*.	Yes, many	IRCEE, PCA genres. Suede-covered hard boards, some pages detached. Emphasis on problem solving.
195	1849–1852	Henry Cort BATCHELDER	M	Phippsburg, Maine	40	Arithmetic 2–	Rs of 3, fractions, interest, barter	Low+	No	No	No	No	In marbled hard cover journal. IRCEE, PCA. Family tree, poems, etc., also in book.
196	1849–1856	Mary Ann DEARDORFF	F	Pennsylvania	198	Arithmetic 1 to 2+	4 opns, Rs of 3, F'ship	Medium	No	No	Yes, Daboll	A few	Disbound, no covers, IRCEE, PCA

United States, Period 9: Cyphering Books Commenced During 1850–1859

M/s ref #	Year/Period	Name(s)	Gender M/F	Location	Number of Pages	Major Content, and Level	Range of Topics: to	Level of Penmanship	Dates on Pages?	Names on Pages?	Linked to Math Text?	Diagrams?	Comments
197	1850	Unknown	Not known	Philadelphia, PA	54	Arithmetic 1	Compound opns, reduction	Medium+	No	No	No	No	Soft-covered, commercial cyphering book.
198	1850–1851	David WISMER	M	Norristown, PA	40	Arithmetic 1	Compound opns	Low +	No	No	No	No	No covers. Transition. PCA but not IRCEE.
199	1850s and 1865–1866	Hilary B. GEARHART, Leidz S. REITZUL	F F	Montgomery, PA	96 (2 m/s)	Arithmetic 1	Compound operations, fractions, reduction, percent	Medium	Yes, from time to time.	No	Yes, but not named	No	No covers, Transition. PCA but not IRCEE. This was being used by Leidz Reitzul in 1866.
200	1851–1852	Philip W. FLORES	M	Upper Milford, Lehigh County, PA	108 (pages numbered)	Mensuration	Finding areas, volumes	High-	Yes, Winter	No	Yes, Bonnycastle	Yes, many.	Hard cover, with leather spine cover.
201	1851–1854	Harriet J. DEARDORFF	M	Pennsylvania	268	Arithmetic 1 to 2	Fractions, interest, R of 3, square root, alligation, f'ship, mensuration	Medium	Yes, Winter	Yes, often	No	No	No covers. Transition. Mainly PCA, no IRCEE. Pages were originally sewn in quires, but have become separated.
202	1854	Herbert TORREY	M	Stirling, NJ	16	Arithmetic 1 to 1+	Fractions, stocks, partial payments.	Medium	No	No	Yes, Greenleaf's National Arithmetic.	No	Entries in pencil. Very practical orientation. PCA but not IRCEE.
203	1855	David DEY	M	Norfolk, Virginia	78	Arithmetic 1	Fractions, decimals, compound opns.	Medium	No	No	Yes, Charles Davies	No	Equal additions for subtraction. Confederate soldier. Still Sterlin currency in 1855. PCA, some IRCEE.

United States, Period 9: Cyphering Books Commenced During 1850–1859 (Continued)

M/s ref #	Year/Period	Name(s)	Gender M/F	Location	Number of Pages	Major Content, and Level	Range of Topics: to …	Level of Penmanship	Dates on Pages?	Names on Pages?	Linked to Math Text?	Diagrams?	Comments
204	1856	Lydia WHITLOCK	F	Location unknown	66	Arithmetic 1	4 opns, compound operations, reduction.	High–	No	No	No	No	Mostly PCA. IRCEE in the Reduction section. Red underlining. Hard covers, leather spine.
205	1856–1858	Jonas BOYER	M	Germantown, Montgomery County, MD	124 pages in 2 sections	Arithmetic 2– to 2+	Rs of 3, to APs and GPs, f'ship	High-(colored pictures)	No	No	No	No	Soft newspaper covers. PCA, but no IRCEE. Transition document.
206	1856–1858	Abraham Isaac GEIMAN	M	Carroll, MD	190 pages (numbered, starting at p. 163)	Arithmetic 1 to 2+	Fractions to Rs of 3 to APs and GPs, permutations	High–	Yes, Winter	Occasionally	No	No	Marbelized hard covers, with leather spine cover. Transition: PCA, not IRCEE.
207	1857	Sarah A. PIERCE	F	Pepperell, MA	36 (2 m/s)	Algebra and Arithmetic 1	Algebra to quadratics fractions	Medium+	No	No	No	No	Rare m/s showing a female doing algebra. At Gilmanton Academy (NH). Transition: PCA, not IRCEE.
208	1857	Josiah URICH	M	Myerstown, PA	41	Arithmetic 1	Fractions, decimals	Medium	No	No	No	No	Transition document. PCA but not IRCEE. Hard covers. Good calligraphic headings.
209	1858–1861	Eddie SCOTT	M	Monticello, Arkansas (1858) and South Gibson, Tennessee (1861)	104	Algebra	Equations, simplifications, etc.	High–	No	No	Linked to Davies' *Algebra*.	No	Transition document. PCA but not IRCEE. Hard covers. Good calligraphic headings.

United States, Period 10: Cyphering Books Commenced During 1860–1869

M/s ref #	Year/Period	Name(s)	Gender M/F	Location	Number of Pages	Major Content, and Level	Range of Topics: to ...	Level of Penmanship	Dates on Pages?	Names on Pages?	Linked to Math Text?	Diagrams?	Comments
210	1860	Thomas Park THOMAS	M	Massachusetts	70	Arithmetic 1 to 2–	4 opns, interest, fractions, mensuration tare and t	High–	Yes, Winter.	No	No	Yes, many beautifully colored figures	Magnificent colored diagrams. IRCEE arises only occasionally. Loose pages between hard covers.
211	c. 1860	John CHAPPELL	M	Maine	24	Arithmetic 2– to 2+	Rs of 3, S and C interest, f'ship	Low–	No	No	No	No	IRCEE, PCA genres. Soft covered commercially-printed, blue-lined book.
212	1860-1861	William F. C. GABLE	M	Lower Windsor Township, York Co., PA	36	Arithmetic 1+	Fractions, interest, discount, partial payments	Medium+	Yes, Winter 1860 and 186.)	No	Questions from Stoddart, Ray, Brooks	No	Transition document. Our youngest cyphering book. PCA but no IRCEE genre.

Appendix B: Cyphering Books Held by the Huguenot Historical Society

(The following is reproduced here with the permission of the Huguenot Historical Society.)

The collection, housed in three boxes, consists of 25 handwritten cyphering books used by students learning mathematics, handwriting, spelling and other disciplines. In many instances, the problems in the cyphering books may have been copied from published contemporary textbooks. Most of the books date from the early 19th century and are bound in cloth, leather, or board. As a whole, the books are in very good condition and quite legible. The majority of the books contain entries in English, although occasional examples of French (#23) and Dutch (#11) are also found.

The collection is an excellent source for documenting the early history of education and children in New Paltz and Ulster County, NY, particularly of descendants of the French and Dutch families who settled New Paltz in the 17th and 18th centuries. Families represented in the books include Chase, Coddington, Deyo, DuBois, Elting, Freer, Hasbrouck, LeFevre, Schoonmaker, Stillwell, and Vreeland. Mathematical content in the books typically includes geometric problems; simple mathematical operations such as addition subtractions, etc.; measurement, problems involving several different currencies; application problems, proportion, interest, decimals, fractions and etc.

Of particular interest are the application problems, which reveal social issues of the times. Religious and moral lessons abound throughout all of the books. Examples include questions and comments such as "He who cannot be happy without great pains will always find his pains greater than his happiness," and "Josiah DuBois is my name, America is my Nation, New Paltz is my dwelling place and Christ is my Salvation. When I am dead and in my grave and all my bones are rotten, when this you see remember me that I am not forgotten." Also of interest are records of financial transactions, genealogical information, names of schoolteachers, and drawings and other evidence of "doodling."

Other cyphering books are located in the Louis Bevier Family Papers: *The Elizabeth Wright Collection (1721–1929)*.

Huguenot Historical Society, 18 Broadhead Avenue, New Paltz, NY 12561, info@huguenotstreet.org, 845-255-1660 (Copyright © 2004 Huguenot Historical Society)

Item List

#1: Josiah LeFevre Cyphering Book (1822–1824)
#2: Stillwell Family Cyphering Book (1829–1842)
#3: Andries J. LeFever Cyphering Book (1799–1811)
#4: Rachel Elting Cyphering Book (1803–1812)
#5: Cornelius D. LeFever Cyphering Book (1820–1825)
#6: Philip Deyo Cyphering Book (1834)
#7: Abraham Deyo Cyphering Book (1804–1846)
#8: Josiah Hasbrouck/Sarah DuBois (1821)
#9: Benjamin Hasbrouck, Jr. Cyphering Book (1764–1766)
#10: Unidentified, "New Paltz School, June 1818, Gilbert Cuthbert Rice preceptor." (1818–1849)
#11: Peter LeFever, Jr. Cyphering Book (1773–1775)
#12: Peter LeFever, Jr. Cyphering Book (1779–1781)
#13: Elias Freer Cyphering Book (1802–1803)
#14: Blandina LeFever Cyphering Book (1833)
#15: Josiah DuBois Cyphering Book (1792–1794)
#16: Hendricus Schoonmaker Cyphering Book (1781–1785)
#17: Merril Chase Cyphering Book (1790–1791, 1836)
#18: Philip Deyo Cyphering Book (1768)
#19: Philip LeFever Cyphering Book (1834)
#20: Jacobus Coddington Cyphering Book (1830)
#21: Jacob Vreeland Cyphering Book (1839)
#22: Johannis Freer Cyphering Book (1797, 1829–1833)
#23: Abraham Hasbrouck Cyphering Book (ca. 1730–1739)
#24: Roelof J. Elting (1823)
#25: Jeremiah A. Houghtaling Cyphering Book (1828)

Item Descriptions

Cyphering Book #1: Josiah LeFevre Cyphering Book (1822–1824). Topics in this book range from simple mathematical operations to more complex problems such as percents and interest and roots of numbers. Other subjects include cloth measure, vulgar fractions, time and weight. In addition to mathematical questions, there are also spelling problems, and application problems involving American History and business and commerce. The handwriting becomes difficult to decipher towards the end of the book.

Cyphering Book #2: Stillwell Family Cyphering Book (1829–1842). This forms part of the *Stillwell/Johnson Family Papers (1789–1943)*. Major topics of study include "reduction," measurement, and direct and indirect proportion. Of interest are numerous "practical questions," or application problems concerning business and commerce. Several of these questions require the student to differentiate the value of the U.S. dollar according to each state's currency. Also found in the book are several "philosophical" quotes such as: "Commendation commonly animates the mind," "He who cannot be happy without great pains will always find his pains greater than his happiness," and "The possession of enjoyment is better than the enjoyment of possession." The book also includes lists of books read and purchased, and their prices. The handwriting in this book is particularly neat and clearly legible.

Cyphering Book #3: Andries J. LeFever Cyphering Book (1799–1811). Includes simple addition, plane geometry, and logarithms. An inserted note reads "Nothing is certain in this world." Other non-mathematical comments include "Be wise and beware," "Command your passions," "Improve your learning," "Beauty commands esteem," and "The children of Mary were twins both boys born June 20, 1811." Other names (or signatures) in the book are Cornelius D. LeFever, Alexander Day, Lewis DuBois, Littyann, Sarah Jane, and Jane. In the book, Andries LeFever notes that he is 18 years old on November 27, 1809.

Cyphering Book #4: Rachel Elting Cyphering Book (1803–1812). Topics range from addition of money to the rule of three. Some pages appear to be missing. Some notes on the bottom of later pages appear to be a record of money saved and spent until 1812. Some application problems were designed to be unique to Rachel. For example, one question is phrased "Suppose Rachel you was born in year of our lord 1792 I desire to know your age to present year 1803 in years, days, hours, and minutes." There are also non-mathematical comments such as "Rachel Elting is my name and so I hope it may remain." A largely illegible note on the inside back cover appears to be addressed to "Mr Solomon Elting my father."

Cyphering Book #5: Cornelius D. LeFever (1820–1825). This leather-bound book contains problems involving units of measure (including motion) and "compound fellowship." The work of Cornelius LeFever is dated from 1820–1825. Also included in the book are: a note for a loan dating from 1859; request for signatures; comments of Catherine Bevier and Jacob Hardenbergh; several stories; a note to a parent from John Clarke. There is also a note from Gertrude Elsie Van Orden DuBois giving Cornelius' age as twelve when he started the book, although his birth year is listed as 1804.

Cyphering Book #6: Philip Deyo Cyphering Book (1834). This was probably donated by Sarah E. Deyo in 1960, and forms part of the *Deyo Family Papers (1675–1870)*. Newspaper was used as binding. Includes sections on simple addition, compound addition, and subtraction, as well as a practice alphabet. Inscription: "Philip Deyo Book and his writing and it is well done and don't steel (sic.) this Book." Also includes a philosophical quote about death.

Cyphering Book #7: Abraham Deyo Cyphering Book (1804–1846). This cloth-covered book was probably donated by Sarah E. Deyo in 1960, and forms part of the *Deyo Family Papers (1675–1870). It* begins with simple addition and ends with simple interest. Also includes a recipe for "Moris Pills," a list of fifteen scripture verses, an 1846 map of "Apel (sic.) Trees," and two pages of cures. There is also an 1830 record of transaction between Abraham Deyo and Elisha Beardsley, and records of birth for sons Jonathan in 1815 and Sallamon (sic.) in 1829.

Cyphering Book #8: Josiah Hasbrouck/Sarah DuBois (1821). This forms part of the *Locust Lawn Collection (1672–1969).* Cloth covered book includes work of both Josiah Hasbrouck and Sarah DuBois. One set of problems is titled "The Elements of Arithmetic, Commenced by Sarah DuBois, Gilbert Cuthbert Ricey, Precepter." Topics include reduction, "promiscuous questions" (applications), vulgar fractions, and domestic exchange. When the book is opened from the back cover, the work of Josiah Hasbrouck is shown. It begins with compound multiplication and ends with inverse proportion. The date January 11, 1821 is included.

Cyphering Book #9: Benjamin Hasbrouck, Jr. Cyphering Book (1764–1766). The questions in this cloth-covered book chiefly involve subtraction, fractions, and interest. Of interest are steps in several problems entitled "Proeve," which involve checking previous work. Some entries are in Dutch. Additional items and comments in the book are the signature of Benjamin Hasbrouck, Sr., dated February 22, 1819; an undated newspaper clipping obituary for Mrs. Solomon Kelder; and an 1865 Huguenot Bank check with the signatures of JM and JJ Hasbrouck.

Cyphering Book #10: Unidentified, "New Paltz School, June 1818, Gilbert Cuthbert Rice, Preceptor" (1818–1849). This book includes simple addition, inverse proportion, interest, and commission. Contains practice exercise such as "How many Barley corns will reach round the globe ... ?" Inside front cover contains payment records from 1836, 1844, and 1849.

Cyphering Book #11: Peter LeFever, Jr. Cyphering Book (1773–1775). This early leather-bound book begins with subtraction of vulgar fractions and also includes topics such as decimals, interest, arithmetical and geometrical progressions, and square and cube roots. A newspaper clipping affixed to front cover dates from 1770. A note dating from January 1775 reads "Behold this year begins like human life with cold and nakedness." Also, one comment reads "Don't steal this Book for fear of shame for look above there stands the owner's name." There is also a note pertaining to lots (property) of Daniel LeFever and Josiah Elting.

Cyphering Book #12: Peter LeFever, Jr. Cyphering Book (1779–1781). This cloth-covered book begins with addition of money and also includes inverse proportion and addition of fractions. Most dates are given as 1781, but there are a few references to 1779. Comments similar to those mentioned in other books are found in the inside froth cover.

Cyphering Book #13: Elias Freer Cyphering Book (1802–1803). This cloth-covered book begins with numerical tables and also includes simple multiplication. There are also several comments and examples of doodling.

Cyphering Book #14: Blandina LeFever Cyphering Book (1833). The decorative title page reads "Miss Blandina LeFever Under the tuition of Miss Sarah Coverly, New Paltz, March 27, 1833." Book is covered in leather. Mathematical content chiefly includes multiplication. Other information includes copies of "A Ballad by Mary Hewitt," and "The Doctor's Song," which includes the name "Abm. D.B. Elting, New Paltz, Ulster County, N.Y."

Cyphering Book #15: Josiah DuBois Cyphering Book (1792–1794). This cloth-covered book includes numeration, measurement, proportion, interest, fractions, decimals, roots, and the Mariner's Compass. The last page identifies Isaiah Plyter, Philomathematician, Dec. 31, 1794. Also in the book is the comment "Josiah DuBois is my name, America is my Nation, New Paltz is my dwelling place and Christ is my Salvation. When I am dead and in my grave and all my bones are rotten, when this you see remember me that I am not forgotten."

Cyphering Book #16: Hendricus Schoonmaker Cyphering Book (1781–1785). The front inside cover of this cloth-covered manuscript contains a newspaper clipping from March 1776. Mathematical content begins with subtraction of measurement and also includes proportion, interest, money exchange, and equation of payments. Smaller pages have been inserted into the book. These pages concern decimals, geometry, trigonometry, and area. A note reads "Henry Schoonmaker has begun a quarter schooling by Mr. Agnew, May 3, 1785." There is also a record of a bond to John Smith, Essex County, NJ, March 20, 1785.

Cyphering Book #17: Merril Chase Cyphering Book (1790–1791, 1836). The title-page reads "Merril Chase Book 1790, Begun at M. Bartlet's School, Dec. 3, 1790." Content includes a numeration table, measurement, and application problems. One sample problem reads "How many days hath elapsed since the birth of Christ to Christmas 1756?" One additional note states "Enoch French began taking papers March 8, 1836."

Cyphering Book #18: Philip Deyo Cyphering Book (1768). Possibly this was donated by Sarah E. Deyo in 1960. It forms part of the *Deyo Family Papers (1675–1870)*. This book contains no cover. Mathematical topics include subtraction of money, interest, measurement, and proportion. The signatures of Jacob and Dinah Elting are present.

Cyphering Book #19: Philip LeFever's Cyphering Book (1834). This paper-covered book contains "Geometrical Problems Preparatory to Surveying." Topics include constructions, right triangles, trigonometry, and area. There are also a significant number of application problems, several of which have elaborate accompanying sketches. There are also some more recent calculations done in pencil.

Cyphering Book #20: Jacobus Coddington's Cyphering Book (1830). This cover-less book is identified as that of Jacobus Coddington of the town of Rochester, NY. Content includes basic operations and measurement, application problems, and a record of accounts.

Cyphering Book #21: Jacob Vreeland's Cyphering Book (1839). This cloth-covered book forms part of the *Vreeland Family Papers (1822–1904)*. Topics include measurement, fractions, decimals, and interest. Additional material concerns an account of a trip from Puerto Rico to Ireland to New York on a ship entitled the "Phoenix."

Cyphering Book #22: Johannes Freer's Cyphering Book (1797, 1829–1833). This book may be three separate books combined in one. Topics are tare and tret, rule of mixture, partnership, bankruptcy, exchange, interest fractions, proportions, and decimals. There are also notes of loans to members of the Freer and DuBois families. The first section dates from 1797 and includes the name of Johannes Freer. The third identifies Johannes J. Freer and dates from 1829–1833.

Cyphering Book #23: Abraham Hasbrouck's Cyphering Book (ca. 1730–1739). Written almost completely in French, this book contains multiplication problems, application problems involving currency (francs, florins, pounds, shillings, etc.), geometric problems, "proeve" applications, and narratives about various mathematical principles and methods. Of particular interest is what appears to be a rough draft of a will. Also included in the book are accounts dating from the 1730s, probably kept by Isaac Hasbrouck, concerning the purchase and sale of books, shirts, pipes, tobacco and other domestic supplies. The names of two schoolteachers, Jean Tebanin and Jean Meschine, are also supplied. The book is wrapped in leather and contains a strap. The front cover contains an inscription and signature by Abraham Hasbrouck.

Cyphering Book #24: Roelof J. Elting (1823). Contains lessons and word problems chiefly concerning business mathematics. Subjects include the double rule of three, interest, equation of payments, barter, loss and gain, fellowship, alligation, exchange, and fractions and decimals. There are also some problems relating to geometry. In addition to the mathematical work, there is one essay entitled "Humanity," which argues that "Humanity is not properly a single virtue; but a disposition residing in the heart, which does not spurn the needy & afflicted, but sends them away relieved from their wants."

Cyphering Book #25: Jeremiah A. Houghtaling (1828–1833). Contains mathematical lessons and word problems, including the rule of three, commission, brokerage, insurance, interest, tare and tret, etc. Of interest are poems and phrases found on the final pages of the book, and numerous dated signatures of Jeremiah A. Houghtaling.

Appendix C: Conversion Tables for US Systems of Measurement, Around 1800

Tables for converting systems of measurement widely used in the United States of America around 1800 are shown. The information is taken from pages 35–44 of the 1822 edition of Daniel Adams, *The Scholar's Arithmetic or Federal Accountant;* ... (Keene, NH: John Prentiss).

1. Money (English currency) *(Adams, p. 35)*

4 Farthings *(qr.)*	make 1 Penny *(d)*
12 Pence	make 1 Shilling *(s)*
20 Shillings	make 1 Pound

Also: 21 shillings make 1 guinea

2. Troy weight *(for gold, silver, jewels, electuaries and liquors) (Adams, p. 37)*

24 grains *(grs.)*	make 1 Pennyweight *(pwt)*
20 Pennyweights	make 1 Ounce *(oz)*
12 Ounces	make 1 Pound *(lb)*

3. Avoirdupois weight *(Adams, p. 38)*

16 drams *(dr.)*	make 1 ounce *(oz)*
16 ounces	make 1 pound *(lb)*
22 pounds	make 1 quarter *(qr)* of a hundred weight *(cwt)*
4 quarters	make 1 hundred weight (or 112 *lbs*)
20 hundred weight	make 1 ton (or 2240 *lbs*)

4. Time *(Adams, p. 39)*

60 seconds *(s)*	make 1 minute *(m)*
60 minutes	make 1 hour *(h)*
24 hours	make 1 day *(d)*
7 days	make 1 week *(w)*
4 weeks	make 1 month *(mo)*
13 *mo*, 1 *d*, and 1 *h*.	make 1 Julian Year *(Y)*

5. Motion *(Adams, p. 40)*

60 seconds	make 1 prime minute
60 minutes	make 1 degree (marked °)
30 degrees	make 1 sign *(s)*
12 signs or 360 degrees	make 1 whole great circle of the zodiac.

6. Cloth measure *(Adams, p. 40)*

2 inches and 1/5 inch	make 1 nail *(na)*
4 nails or 9 inches	make 1 quarter of a yard *(qr)*
4 quarters of a yard or 36 inches	make 1 yard *(yd)*
3 quarters of a yard, or 27 inches	make 1 ell *(E. Fl.)* Flemish
5 quarters of a yard, or 45 inches	make 1 ell *(E. E.)* English
6 quarters of a yard, or 54 Inches	make 1 ell *(E. Fr.)* French

7. Long measure *(Adams, p. 41)*

3 barley corns (*bar.*)	make 1 inch (*in*)
12 inches	make 1 foot (*ft*)
3 feet	make 1 yard (*yd*)
$5\frac{1}{2}$ yards or $16\frac{1}{2}$ ft	make 1 rod, perch, or pole (*pol*)
40 poles	make 1 furlong (*fur*)
8 furlongs	make 1 mile (*mile*)
$69\frac{1}{2}$ statute miles	make 1 degree of a great circle
360 degrees	Make a great circle of the earth

4 quarters 1 inch and 1 fifth, or 37 inches and one fifth	make 1 ell (*E. sc.*) Scotch
3 quarters and two thirds	make 1 Spanish var

8. Land or square measure *(Adams, p. 42)**

144 inches	make 1 square foot
9 feet	make 1 square yard
$30\frac{1}{4}$ yards or $272\frac{1}{4}$ feet	make 1 pole
40 poles	make 1 rood
4 roods 160 rods or 4340 yards	make 1 acre
640 acres	make 1 mile

* Note that Adams did not say 144 square inches made 1 square foot, or 9 square feet made 1 square yard, etc. The reason why Adams (and other writers) asserted "144 inches makes 1 sq ft" is unclear.

9. Solid measure *(Adams, p. 42)*

1728 inches	make 1 foot
27 feet	make 1 yard

40 feet of round timber or 30 feet of hewn timber, make 1 ton or load.

128 solid feet, i.e., 8 inches long, 4 inches wide, and 4 inches high, make 1 cord of wood.

Other relationships (Adams, p. 45)
12 particular things make 1 dozen.
12 dozen things make 1 gross.
12 gross or 144 dozen make 1 great gross.
Also, 20 particular things make 1 score.

10. Wine measure *(Adams, p. 43)*

2 pints (*pts*)	make 1 quart (*qt*)
4 quarts	make 1 gallon (*gal*)
10 gallons	make 1 anchor of brandy (*anc*)
13 gallons	make 1 runlet (*run*)
$31\frac{1}{2}$ gallons	make half hogshead ($\frac{1}{2}$*hhd*)
42 gallons	make 1 tierce (*tier*)
63 gallons	make 1 hogshead (*hhd*)
2 hogsheads	make 1 pipe or butt (*P* or *B*)
2 pipes	make 1 tun (*T*)

6 points make 1 line
12 lines make 1 inch
4 inches make 1 hand for measuring horses)
3 hands make 1 foot
66 feet, or 4 poles, make a Gunter's chain
3 miles make 1 league (distances at sea)
1 quintal of fish weighs 1 cwt., avoirdupois.

11. Ale or beer measure *(Adams, p. 44)*

2 pints	make 1 quart (*qt*)
4 quarts	make 1 gallon (*gal*)
8 gallons	make 1 firkin of ale in London (*fir*)
$8\frac{1}{2}$ gallons	make 1 firkin of ale or beer (*fir*)

12. Dry measure *(Adams, p. 44)*

2 pints	make 1 quart (*qt*)
2 quarts	make 1 pottle (*pot*)
2 pottles	make 1 gallon (*gal*)
2 gallons	make 1 peck (*pk*)

Appendix C: Conversion Tables for US Systems of Measurement, Around 1800   193

9 gallons	make 1 firkin of beer in London (fir)	4 pecks	make 1 bushel (*bush*)
2 firkins	make 1 kilderkin (*kild*)	2 bushels	make 1 strike (*str*)
2 kilderkins	make 1 barrel (*bar*)	2 strikes	make 1 coom (*co*)
1½ barrels, or 54 gallons	make 1 hogshead of beer (*hhd*)	2 cooms	make 1 quarter (*qr*)
2 barrels	make 1 puncheon (*pun*)	4 quarters	make 1 chaldron (*ch*)
3 barrels, or 2 hogsheads	make 1 butt (*butt*)	4½ quarters	make 1 chaldron (in London)
		5 quarters	make 1 wey (*wey*)
		2 weys	make 1 last (*last*)

References

(Note: Cyphering books in the principal data set are not listed here—Each of those cyphering books is briefly summarized in Appendix A)

A Carthusian. (1847). *Chronicles of Charterhouse*. London, UK: George Bell.
A Catechism of Arithmetic, Containing a Concise Explanation of the most Useful Rules; with a Variety of Questions and Explanatory Notes, and Exercises in each Rule (1821) (6th ed.). London, UK: W. B. Whittaker.
A General View of the Conduct of the French in America and of our Settlements there. (1745). *Gentleman's Magazine* (London, UK), 25, 15.
Ackerberg-Hastings, A. (2000). *Mathematics is a gentleman's art: Analysis and synthesis in American college geometry teaching 1790–1840*. PhD dissertation, Iowa State University.
Ackerberg-Hastings, A. (2002). Analysis and synthesis in John Playfair's "Elements of Geometry." *The British Journal of History of Science, 35*, 43–72.
Ackerberg-Hastings, A. (2009, June). *The relationship between mathematics and physical sciences in John Playfair's Natural Philosophy course*. Paper presented at the annual conference of the Canadian Society for History and Philosophy of Mathematics, held at St John's, Newfoundland.
Adams, D. (1801). *The scholar's arithmetic—Or Federal accountant*. Leominster, MA: Adams and Wilder.
Adams, D. (1822). *The scholar's arithmetic—Or Federal accountant*. Keene, NH: John Prentiss.
Adams, D. (1827). *Adams's new arithmetic: Arithmetic in which the principles of operating by numbers are analytically explained and synthetically explained, thus combining the advantages to be derived both from the inductive and synthetic mode of instructing*. Keene, NH: J. Prentiss.
Adams, O. F. (1903). *Some famous American schools*. Boston, MA: Dana Estes & Company.
Alexander, J. (1709). *A synopsis of algebra*. London, UK: J. Barber.
Allan, G. A. T. (1984). *Christ's Hospital*. London, UK: Town & Country Books.
Allen, J. B. L. (1970). *The English mathematical schools, 1670–1720*. PhD thesis, University of Reading.
Allison, R. J. (Ed.). (1997). *American eras: The Reform era and eastern United States development, 1815–1850* (Vol. 5). Detroit, MI: Gale.
Ambulator. (1780, September). To the man of pleasure. *The Town and Country Magazine, or Universal Repository of Knowledge, Instruction and Entertainment, 478*.
Ames, G. J. (2008). *The globe encompassed: The age of European discovery, 1500–1700*. Lebanon, IN: Pearson Prentice Hall.
Ames, S. M. (1958). *Reading, writing and arithmetic in Virginia, 1607–1699*. Williamsburg, VA: Virginia 350th Anniversary Celebration Corporation.
Andrews, C. C. (1830). *The history of the New York African free schools, from their establishment in 1787, to the present time*. New York, NY: M. Day.
Andrews, C. M. (1912). *The colonial period*. New York, NY: Henry Holt and Company.

Angus, D. L., Mirel, J. E., & Vinovskis, M. A. (1988). Historical development of age stratification in schooling. *Teachers College Record, 90*(2), 211–236.
Archer, L. J. (Ed.). (1988). *Slavery and other forms of unfree labor*. London, UK: Routledge.
Arrighi, G. (Ed.). (1964). *Paolo dell'abbaco. Trattato d'aritmetica*. Pisa, Italy: Domus Galileana.
Arrighi, G. (Ed.). (1970). *Piero della Francesca: Trattato de abaco*. Pisa, Italy: Domus Galileana.
Ayres, J. (1682). *Arithmetick and writing*. London, UK: Author.
Babu, D. S. (2007). Memory and mathematics in the Tamil Tinnai schools of South India in the 18th and 19th centuries. *International Journal for the History of Mathematics Education, 2*(1), 15–37.
Bache, A. D. (1839). *Report on education in Europe, to the trustees of the Girard College for Orphans*. Philadelphia, PA: Bailey.
Bailyn, B. (1960). *Education in the forming of American society: Needs and opportunities for study*. New York, NY: W. W. Norton & Company.
Baker, H. (1568/1591). *The well-spring of sciences, which teacheth the perfect worke and practice of arithmeticke, both in whole numbers and in fractions*. London, UK: Thomas Purfoot.
Baker, H. (1687). *Baker's arithmetick teaching the perfect worke and practice of arithmeticke, both in whole numbers and in fractions, whereunto are added many rules and tables of interest, rebate and purchases, &c, also the art of decimal fractions intermixed with common fractions for the better understanding thereof, newly corrected and contracted and made more plain and ease by Henry Phillippes*. London, UK: J. Richardson and William Thackery.
Bangs, J. D. (2000). Pilgrim homes in Leiden. *The New England Historical and Genealogical Register, 154*(616), 413–445.
Barnard, H. (1851). *Normal schools, and other institutions, agencies and means designed for the professional education of teachers*. Hartford, CT: Case, Tiffany and Company.
Barnard, H. (1856). Graduation of public schools with special reference to cities and large villages. *American Journal of Education, 2*, 455–464.
Barnard, H. (Ed.). (1859). *Life, educational principles, and methods of John Henry Pestalozzi, with biographical sketches of several of his assistants and disciples*. New York, NY: F. C. Brownell.
Barrème, N. (1747). *L'Arithmétique du Sieur de Barrème ou le livre facile pour apprendre l' arithmétique de soi-meme et sans maître*. Paris, France: Gandouin.
Bartlett, J. R. (1832/1933). *Letter of instructions to the captain and the supercargo of the brig "Agenoria," engaged in a trading voyage to Africa*. Philadelphia, PA: Howard Greene and Arnold Talbot.
Bean, S. S. (2001). *Yankee India: American commercial and cultural encounters with India in the age of sail 1784–1860*. Salem, MA: Peabody Essex Museum.
Beaujouan, G. (1988). The place of Nicolas Chuquet in a typology of 15th century French arithmetics. In C. Hay (Ed.), *Mathematics from manuscript to print 1300–1600* (pp. 73–88). Oxford, UK: Clarendon Press.
Beckers, D. J. (1999). *"Come children!" Some changes in Dutch arithmetic books 1750–1850*. Report 9902 of the Department of Mathematics, the University of Nijmegen.
Bellhouse, D. (2005, April). *Probability and statistics ideas in the classroom—Lessons from history*. Paper presented at the 55th Session of the International Statistical Institute, held in Sydney, Australia.
Bellhouse, D. (2010, May). *The mathematics curriculum in the British dissenting academies in the 18th century*. Paper presented at the meeting of the Canadian Society for the History and Philosophy of Mathematics, held in Montréal.
Bennett, T. (1815). *New system of practical arithmetic peculiarly calculated for the use of schools in the United States* (5th ed.). Philadelphia, PA: Bennett & Walton.
Benoit, P. (1988). The commercial arithmetic of Nicolas Chuquet. In C. Hay (Ed.), *Mathematics from manuscript to print 1300–1600* (pp. 96–116). Oxford, UK: Clarendon Press.
Best, J. H. (Ed.). (1967/1962). *Benjamin Franklin on education*. New York, NY: Teachers College Press.

Bethlehem Digital History Project. (2009). *Susan Shimer cyphering book.* Retrieved September 27, 2009, from http://bdhp.moravian. edu/ bethlehem/bethlehem.html

Bickham, G. (1740). *Youth's instructor in the art of numbers, a new cyphering book in which is shewn variety of penmanship, command of hand: Engraved for the use of schools.* London, UK: William & Cluer Dicey.

Bjarnadóttir, K. (2009, October 15). *Arithmetica—Arithmetic art, mathematics for all.* Paper presented at the Nordic Research Network on Special Needs Education in Mathematics (NORSMA) Conference, held at the University of Iceland.

Bonnycastle, J. (1788). *The scholar's guide to arithmetic; Or a complete exercise-book for the use of schools, with notes containing the reason of every rule, . . .* (5th ed.). London, UK: J. Johnson.

Boston Schoolmasters. (1844). *Remarks on the Seventh Annual Report of the Hon. Horace Mann, Secretary of the Massachusetts Board of Education.* Boston, MA: Charles C. Little and James Brown.

Bowditch, N. (1797). *Journal of a voyage from Salem to Manila in the ship Astrea, E. Prince, Master, in the years 1796 and 1797.* Handwritten manuscript held in the Bowditch Collection, Boston Public Library.

Boyden, A. C. (1933). *History of Bridgewater Normal School.* Bridgewater, MA: Bridgewater Normal Alumni Association.

Brayley, A. W. (1894). *Schools and schoolboys of old Boston.* Boston, MA: Louis P. Hager.

Breed, F. S., Overman, J. R., & Woody, C. (1936). *Child life arithmetics. Grade Five.* Chicago, IL: Lyons & Carnahan.

Bregman, A. (2005). Alligation alternate and the composition of medicines: Arithmetic and medicine in early modern England. *Medical History, 49*(3), 299–320.

Bremner, R. H. (1970). *Children and youth in America: A documentary history, Volume 1, 1600–1865.* London, UK: Oxford University Press.

Briggs, A., & Burke, P. (2010). *A social history of the media: From Gutenberg to the Internet* (3rd ed.). Malden, MA: Polity Press.

Brockliss, L. W. B. (1987). *French higher education in the seventeenth and 18th centuries—A cultural history.* New York, NY: Oxford University Press.

Brooks, B. C. (2010). *Dethroning the Kings of Cape Fear: Consequences of Edward Moseley's surveys.* Unpublished History 4000 thesis, East Carolina University.

Brooks, E. (1860). *Methods of teaching mental arithmetic and key to the normal mental arithmetic.* Philadelphia, PA: Sower, Barnes & Co.

Brooks, E. (1876). *Philosophy of arithmetic as developed from the three fundamental processes of synthesis, analysis, and comparison containing also a history of arithmetic.* Philadelphia, PA: Sower, Potts & Company.

Brooks, E. (1879). *Normal methods of teaching containing a brief statement of the principles and methods of the science and art of teaching.* Philadelphia, PA: Normal Publishers.

Broome, E. C. (1903). *A historical and critical discussion of college-entrance admissions requirements.* New York, NY: Columbia University.

Brown, J. C., & Coffman, L. D. (1914). *How to teach arithmetic: A manual for teachers and a text-book for normal schools.* Chicago, IL: Row Peterson and Company.

Brubacher, J. S., & Rudy, W. (1958). *Higher education in transition: A history of American colleges and universities, 1636–1968.* New York, NY: Harper and Row.

Bureau of Education. (1905). *Report of the Commissioner of Education for the Year 1903.* Washington, DC: Author.

Burke, E., & Burke, W. (1758). *An account of the European settlements in America* (Vol. 2). London, UK: R. and J. Dodsley.

Burrowes, T. H. (1862). Twenty-eighth annual report of the Superintendent of Schools of Pennsylvania. *The Pennsylvania School Journal, February 1862,* 243–256.

Burton, W. (1833). *The district school as it was, by one who went to it.* Boston, MA: Carter, Hendee and Company.

Burtt, E. H., & Davis, W. E. (1995). Historic and taxonomic implications of recently found artwork in arithmetic books of students of Alexander Wilson. *The Wilson Bulletin, 107*(2), 193–215.

Butler, W. (1788). *An introduction to arithmetic designed for the use of young ladies.* London, UK: S. Couchman.

Butler, W. (1788/1806). *Arithmetical questions, on a new plan, intended to answer the double purpose of arithmetical instruction and miscellaneous information, to which are subjoined, a collection of arithmetical tables, questions for practical examination and a copious index of persons, places and things, occasionally treated of, or mentioned in the work, designed for the use of young ladies* (4th ed.). London, UK: S. Couchman.

Butler, W. (1819). *An introduction to arithmetic designed for the use of young ladies* (3rd ed.). London, UK: Simpkin, Marshall & Co.

Cajori, F. (1890). *The teaching and history of mathematics in the United States* (Circular of Information No. 3, 1890). Washington, DC: Bureau of Education.

Cajori, F. (1907). *A history of elementary mathematics with hints on methods of teaching.* New York, NY: The Macmillan Company.

Caldwell, O. W., & Courtis, S. A. (1925). *Then and now in education, 1845–1923.* Yonkers-on-Hudson: New York, NY: World Book Company.

Campbell, D. (1892). *The puritan in Holland, England, and America.* New York, NY: Harper & Brothers Publishers.

Carlo, P. W. (2005). *Huguenot refugees in colonial New York: Becoming American in the Hudson Valley.* Brighton, UK: Sussex Academic Press.

Carpenter, C. (1963). *History of American schoolbooks.* Philadelphia, PA: University of Pennsylvania Press.

Chambers' Information for the People. (1868). *Mechanics' institutions* (Vol. 1, pp. 713–744). New York, NY: United States Publishing Company.

Chambers, R. (1835). *Cyphering book.* Bethania, PA: Author.

Champion, R. J. (1747). *The tutor's assistant in teaching arithmetic.* London, UK: Author.

Child, J. (1693). *A new discourse of trade.* London, UK: John Everingham.

Christ's Hospital. (1595). *Dame Mary Ramsey's gift: Deeds relating to the maintenance by the Governors of Christ's Hospital of a grammar school in Halstead, Essex.* Reference Code CLC/210/G/BRB/041/MS13583. (Document held in Guildhall, London, UK).

Christ's Hospital. (1785). *Changes and orders for the several officers of Christ's Hospital reviewed by the Committee of Almoners at several meetings; And approved and confirmed by general courts held in the said Hospital . . .in 1784 and 1785.* London, UK: Charles Rivington.

Christ's Hospital. (1857). *Register of Christ's Hospital Royal Mathematical School boys apprenticed chips' masters 1816–1857.* Handwritten manuscript, London Metropolitan Archives Reference Code CLC/526/MS30338.

Christ's Hospital. (1953). *The Christ's Hospital book.* London, UK: Hamish Hamilton.

Clarke, J. F., & Hale, E. E. (1892). School days in New England. In K. Munroe & M. H. Catherwood (Eds.), *School and college days* (Vol. VII, pp. 265–280). Boston, MA: Hall and Locke Company.

Clason, R. (1968). *Number ideas of arithmetic texts of the United States from 1880 to 1966 with related psychological and mathematical developments.* PhD dissertation, University of Michigan.

Clemens, S. (Mark Twain). (1885). *Adventures of Huckleberry Finn.* New York, NY: Charles L. Webster.

Clements, M. A., & Ellerton, N. F. (1996). *Mathematics education research: Past, present and future.* Bangkok, Thailand: UNESCO.

Cobb, L. (1834). *Cobb's explanatory arithmetick, number two; Containing the compound rules, and all that is necessary of every other rule in arithmetick for practical purposes and the transactions of business; Adapted to the understanding and use of larger children in schools and academies. To which is annexed a practical system of book-keeping.* Elmira, NY: Birdsall & Huntley.

References

Cobb, L. (1835). *Cobb's cyphering book, No. 1, containing all the sums and questions for theoretical and practical exercises in Cobb's Explanatory Arithmetic No 1*. Elmira, NY: Birdsall & Huntley.

Cockburn, J. S., King, H. P. F., & McDonnell, K. G. T. (Eds.). (1969). Private education from the sixteenth century: Developments from the 16th to the early 19th century. *A History of the County of Middlesex: Volume 1: Physique, archaeology, domesday, ecclesiastical organization, the Jews, religious houses, education of working classes to 1870, Private Education from Sixteenth Century* (pp. 241–255). Retrieved February 10, 2011, from http://www.british-history.ac.uk/report.aspx?compid=22124

Cocker, E. (1664). *The tutor to writing and arithmetic*. London, UK: H. Tracy.

Cocker, E. (1678/1725). *Cocker's arithmetick: A plain and familiar method . . .* (42nd ed.). London, UK: H. Tracy.

Cocker, E. (1684/1720). *Decimal arithmetic, wherein is shewed the nature and use of decimal fractions in the usual rules of arithmetic, . . .* (5th ed.). London, UK: J. Darby for M. Wellington.

Cohen, P. C. (1982). *A calculating people: The spread of numeracy in early America*. Chicago, IL: University of Chicago Press.

Cohen, P. C. (1993). Reckoning with commerce: Numeracy in 18th-century America. In J. Brewer & R. Porter (Eds.), *Consumption and the world of goods* (pp. 320–334). London, UK: Routledge.

Cohen, P. C. (2003). Numeracy in 19th-century America. In G. M. A. Stanic & J. Kilpatrick (Eds.), *A history of school mathematics* (Vol. 1, pp. 43–76). Reston, VA: National Council of Teachers of Mathematics.

Colburn, D. P. (1855). *Arithmetic and its applications: Designed as a text book for common schools, high schools, and academies*. Philadelphia, PA: H. Copperthwait.

Colburn, W. (1821). *An arithmetic on the plan of Pestalozzi, with some improvements*. Boston, MA: Cummings and Hilliard.

Colburn, W. (1822). *Arithmetic upon the inductive method of instruction being a sequel to intellectual arithmetic*. Boston, MA: Cummings and Hilliard.

Colburn, W. (1830/1970). Teaching of arithmetic. In J. K. Bidwell & R. G. Clason (Eds.), *Readings in the history of mathematics education* (pp. 24–37). Washington, DC: National Council of Teachers of Mathematics.

Coldham, P. W. (1990). *Child apprentices in America from Christ's Hospital, London, 1617–1778*. Baltimore, MD: Genealogical Publishers Co.

Collinder, P. (1954). *A history of marine navigation*. London, UK: B. T. Batsford Ltd.

Colson, J. (1736.). *An arithmetical copy-book*. London, UK: Author.

Connor, R. (1902). *Glengarry school days: A story of early days in Glengarry*. New York, NY: Grosset & Dunlap Publishers.

Coolidge, O. (1974). *The apprenticeship of Abraham Lincoln*. New York, NY: Charles Scribner's Sons.

Coon, C. L. (1915). *North Carolina schools and academies 1790–1840: A documentary history*. Raleigh, NC: Edwards and Broughton Printing Co.

Cornell, E. (1860). *Cyphering book prepared In 1823 and 1824, with later comments inserted by Ezra Cornell*. Ithaca, NY: Cornell University Archives.

Cowley, E. (1923). An Italian mathematical manuscript. In C. F. Fiske (Ed.), *Vassar medieval studies* (pp. 379–405). New Haven, CT.

Cracraft, J. (1971). *The church reform of Peter the Great*. London, UK: The Macmillan Company.

Craik, A. D. D. (2007). *Mr Hopkins' men*. London, UK: Springer.

Cremin, L. A. (1970). *American education: The colonial experience 1607–1783*. New York, NY: Harper & Row.

Cremin, L. A. (1977). *Traditions of American education*. New York, NY: Basic Books.

Cross, A. (2007). *By the banks of the Neva*. Cambridge, UK: Cambridge University Press.

Cubberley, E. P. (1920). *The history of education*. Boston, MA: Houghton Mifflin.

Cubberley, E. P. (1962). *Public education in the United States*. Boston, MA: Houghton Mifflin.

Curbera, G. P. (2009). *Mathematicians of the world, unite! The International Congress of Mathematicians—A human endeavour*. Wellesley, MA: A. K. Peters Ltd.

Daboll, N. (1800). *Daboll's schoolmaster's assistant: Being a plain, practical system of arithmetic; adapted to the United States*. New London, CT: Samuel Green.

D'Ambrosio, U. (1986). Socio-cultural bases for mathematical education. In M. Carss (Ed.), *Proceedings of the fifth international Congress on Mathematical Education* (pp. 1–6). Boston, MA: Birkhauser.

Dash, J. (2000). *The longitude prize*. New York, NY: Frances Foster Books.

da Silva, M. C. L., & Valente, W. R. (2009). Students' notebooks as a source of research on the history of mathematics education. *International Journal for the History of Mathematics Education, 4*(1), 51–64.

Davenport, B. (1832). *A new gazetteer, or geographical dictionary of North America and the West Indies . . . .* Baltimore, MD: George McDowell, & Son.

Davies, C. (1870). *University arithmetic: Embracing the science of numbers, and general rules for their application*. New York, NY: A. S. Barnes and Co.

Davis, N. Z. (1960). Sixteenth-century French arithmetics on the business life. *Journal of History of Ideas, 21*(1), 18–48.

Day, J. (1814). *An introduction to algebra, being the first part of a course of mathematics, adapted to the method of instruction in the American colleges*. New Haven, CT: Howe and DeForest.

DeGarmo, C. (1896). *Herbart and the Herbartians*. New York, NY: Charles Scribner's Sons.

De Morgan, A. (1847). *Arithmetical books from the invention of printing to the present time*. London, UK: Taylor and Walton.

De Morgan, A. (1853). *The elements of arithmetic*. London, UK: J. Walton.

Dewalt, M. W. (2006). *Amish education in the United States and Canada*. Lanham, MD: Rowman & Littlefield Education.

Dickens, C. (1850). *David Copperfield*. London, UK: Bradbury & Evans.

Dilworth, T. (1797). *Dilworth's arithmetic, being a compendium of arithmetic with both practical and theoretical*. New London, CT: Napthali Judah.

Ditton, H. (1709). *An appendix to the algebra of John Alexander*. London, UK: J. Barber.

Dix, D. L. (1824/1826). *Conversations on common things; Or, guide to knowledge, with questions*. Boston, MA: Munroe and Francis.

Dixon, T. (1630). (Handwritten). *Arithmetic cyphering book*. New York, NY: Plimpton Collection, Butler Library, Columbia University. (Plimpton Collection, Post 1600, MS 510).

Doar, A. K. (2006). *Cipher books in the Southern Historical Collection*. Master of Science thesis, Wilson Library, University of North Carolina at Chapel Hill.

Dodd, J. B. (1856). *An essay on prevailing systems of instruction in elementary mathematics*. New York, NY: Farmer, Brace & Co.

Donoghue, E. F. (2003a). Algebra and geometry textbooks in 20th-century America. In G. M. A. Stanic & J. Kilpatrick (Eds.), *A history of school mathematics* (Vol. 1, pp. 329–398). Reston, VA: NCTM.

Donoghue, E. F. (2003b). The emergence of a profession: Mathematics education in the United States, 1890–1920. In G. M. A. Stanic & J. Kilpatrick (Eds.), *A history of school mathematics* (Vol. 1, pp. 159–193). Reston, VA: NCTM.

Dossey, J, Halvorsen, K., & McCrone, S. (2008). *Mathematics education in the United States 2008*. Reston, VA: NCTM.

Douglas, P. H. (1921). *American apprenticeships and industrial education*. Ph.D. dissertation, Columbia University.

Dowling, D. (1829). *New and improved system of calculation in which a universal rule of proportion is, by new arrangement, applied to questions relating to military affairs, mensuration, natural philosophy, and mercantile operations*. London, UK: Author.

Durkin, J. J. (1942). Journal of the Revd. Adam Marshall, schoolmaster, *U.S.S. North Carolina*, 1824–1825. *Records of the American Catholic Historical Society of Philadelphia, 53*(4), 152–168.

Dyer, E. (1886). The old schools of Providence. *Narragansett Historical Register, 5*(1), 220–240.
Earle, A. M. (1899). *Child-life in colonial days*. New York, NY: The Macmillan Company.
Eby, F., & Arrowood, C. F. (1934). *The development of modern education: In theory, organization, and practice*. New York, NY: Prentice-Hall.
Edwards, R. (1857). *Memoir of Nicholas Tillinghast, first Principal of the State Normal School at Bridgewater, Massachusetts*. Boston, MA: James Robinson & Co.
Eggleston, E. (1871). *The Hoosier schoolmaster*. New York, NY: Orange Judd.
Ellerton, N. F., & Clements, M. A. (2008). An opportunity lost in the history of school mathematics: Noah Webster and Nicolas Pike. In O. Figueras, J. L. Cortina, S. Alatorre, & A. Meplveda (Eds.), *Proceedings of the Joint Meeting of PME 32 and PME-NA XXX* (Vol. 2, pp. 447–454). Morelia, Mexico: Cinvestav-UMSWH.
Ellerton, N. F., & Clements, M. A. (2009a). *Summary of cyphering books at the Phillips Library, Salem, Massachusetts*. Salem, MA: Phillips Library at the Peabody Essex Museum.
Ellerton, N. F., & Clements, M. A. (2009b). Theoretical bases implicit in the abbaci and cyphering-book traditions. In M. Tzekaki, M. Kaldrimidou, & H. Sakonidis (Eds.), *Proceedings of the 33rd conference of the International Group for the Psychology of Mathematics Education* (Vol. 3, pp. 9–16). Thessaloniki, Greece: International Group for the Psychology of Mathematics Education.
Emerson, F. (1832). *The North American arithmetic: Part second, uniting oral and written exercises in corresponding chapters*. Boston, MA: Lincoln and Edmands.
Emerson, F. (1835). *The North American arithmetic: Part third, for advanced scholars*. Boston, MA: Russell, Odiorne & Metcalf.
Emigh, R. J. (2002). Numeracy or enumeration? The uses of numbers by states and societies. *Social Science History, 26*(4), 653–698.
Ewing, A. (1799). *A synopsis of practical mathematics: Containing plane trigonometry, mensuration of heights, distances, surfaces and solids, gauging, surveying of land, navigation and gunnery* (4th ed.). London, UK: T. Cadlell & W. Davies.
Falola, T., & Warnock, A. (Eds.). (2007). *Encyclopedia of the middle passage: Milestones in African American history*. Westport, CT: Greenwood Publishing Group.
Fayazmanesh, S. (2006). *Money and exchange: Folk tales and reality*. New York, NY: Routledge.
Fink, K. (1900). *A brief history of mathematics: An authorized translation of Dr Karl Fink's Geschichte der elementar-mathematik* (W. W. Beman & D. E. Smith, Trans.). Chicago, IL: Open Court.
Finkelstein, B. (1989). *Governing the young: Teacher behavior in popular and primary schools in 19th century United States*. London, UK: Falmer Press.
Flibbert, J., Goss, K. D., McAllister, J., Tolles, B. F., & Trask, R. B. (1999). *Salem: Cornerstones of a historic city*. Beverly, MA: Commonwealth Editions.
Fowle, W. B. (1866). *The teacher's institute, or familiar hints to young teachers*. New York, NY: A. S. Barnes.
Fox, J. (1809). *Hints to the managers and committees of charity and Sunday schools on the practicality of extending such instruction upon Mr Lancaster's plan*. London, UK: Dartmead.
Franci, R. (1992). Le matamatiche dell'abaco nel quatrocento. In *Contributi alla storia delle matematiche: Scritti in onore di Gino Arrighi* (pp. 53–74). Modena, Italy: Mucchi.
Franci, R. (2009). Mathematics in the manuscript of Michael of Rhodes. In P. O. Long, D. McGee, & A. M. Stahl (Eds.), *The book of Michael of Rhodes: A 15th maritime manuscript* (Vol. 3, pp. 115–146). Cambridge, MA: MIT Press.
Franci, R., & Rigatelli, L. T. (1982). *Introduzione all'aritmetica mercantile del medioevo e del rinascimento. Realizzata attraverso un'antologis degli scritti di Dionigi Gori* (Sec VVI). Urbino, Italy: Quattro Venti.
Franci, R., & Rigatelli, L. T. (1988). 14th Italian algebra. In C. Hay (Ed.), *Mathematics from manuscript to print 1300–1600* (pp. 11–29). Oxford, UK: Clarendon Press.

Franci, R., & Rigatelli, L. T. (1989). La matematica nella tradizione dell'abaco nel XIV e XV secolo. In P. Freguglia & C. Maccagni (Eds.), La storia dell scienze (Vol. 5, Book 2) of *Storia sociale e culturale d'Italia* (pp. 68–94). Busto Arsizio, Italy: Bramante Editrice.

Freeman, F. N. (1913). Writing. In P. Monroe (Ed.), *A cyclopedia of education* (Vol. 5, pp. 819–827). New York, NY: The Macmillan Company.

French, J. H. (1869). *Common school arithmetic combining the elements of science with the practical applications to business.* New York, NY: Harper & Brothers.

Fuess, C. M. (1917). *A history of Phillips Academy.* Boston, MA: Houghton Mifflin Company.

Gay, E. (1806). (Handwritten) *Algebra cyphering book.* New York, NY: Plimpton Collection, Butler Library, Columbia University (Plimpton Manuscript MS 511, 1806).

Gay, E. (1812). (Handwritten) *Trigonometry and navigation cyphering book.* New York, NY: Plimpton Collection, Butler Library, Columbia University. (Plimpton Manuscript MS 511, 1806).

Gay, E. (1820). (Handwritten) *Diary of the Reverend Ebenezer Gay, 1810–1824.* New York, NY: Plimpton Collection, Butler Library, Columbia University. Plimpton Manuscript MS 920, 1806).

Gaydos, T., & Kampas, B. (2010). *American and Canadian cyphering books, n.d., 1727–1864* (2nd ed.). Salem, MA: Phillips Library at the Peabody Essex Museum.

Gibbs, G. (1931). *Supercargo.* New York, NY: D. Appleton and Company.

Gies, J., & Gies, F. (1969). *Leonard of Pisa and the new mathematics of the Middle Ages.* New York, NY: Thomas Y. Crowell.

Gilman, C. (1852). *Recollections of a New England bride and of a Southern matron.* New York, NY: G. P. Putnam.

Glaeser, G. (1984). *Racines historiques de la didactique des mathématiques.* Strasbourg, France: IREM de Strasbourg.

Goodrich, S. G. (1833). *Peter Parley's method of teaching arithmetic to children.* Boston, MA: Carter and Hendee.

Goodrich, S. G. (1857). *Recollections of a lifetime, or men and things I have seen: In a series of familiar letters to a friend, historical, biographical, anecdotal and descriptive.* New York, NY: Miller, Orton & Co.

Greenwood, I. (1729). *Arithmetick, vulgar and decimal, with the application thereof to a variety of cases in trade and commerce.* Boston, MA: Kneeland & Green.

Grendler, P. F. (1989). *Schooling in Renaissance Italy.* Baltimore, MD: Johns Hopkins University Press.

Griffis, W. E. (1909). *The story of New Netherland.* Boston, MA: The Riverside Press.

Griffith, P. (1976). *French artillery.* London, UK: Almark.

Grove, M. (2000). *Legacy of one-room schools.* Morgantown, PA: Masthof Press.

Guralnick, S. (1975). *Science and the antebellum American college.* Philadelphia, PA: American Philosophical Spociety.

Halwas, R. (1997). *American mathematics textbooks 1760–1850.* London, UK: Author.

Hans, N. A. (1951a). *New trends in education in the 18th century.* London, UK: Routledge.

Hans, N. A. (1951b). The Moscow School of Mathematics and Navigation (1701). *The Slavonic and East European Review, 29*(73), 532–536.

Harper, C. (1935). *Development of the teachers college in the United States with special reference to Illinois State University.* Bloomington, IL: McKnight & McKnight.

Harper, C. (1939). *A century of public teacher education: The story of the state teachers colleges as they evolved from the normal schools.* Washington, DC: American Association of Teachers Colleges.

Hay, C. (Ed.). (1988). *Mathematics from manuscript to print 1300–1600.* Oxford, UK: Clarendon Press.

Heal, A. (1931). *The English writing-masters and their copy-books 1570–1800.* Cambridge, UK: Cambridge University Press.

Heeffer, A. (2008). Text production reproduction and appropriation within the abbaco tradition: A case study. *SCIAMVS, 9*, 101–145.

Hendrick, E. (1810). *A new and plain system of arithmetic, containing the several rules of that useful science, concisely defined and greatly simplified. The whole, particularly adapted to the easy and regular instruction of youth and to the trade and commerce of the United States.* Richmond, VA: Lynch and Davis.

Henry, J. (1843). *An address upon education and common schools.* Albany, NY: Staem Press.

Herttenstein, J. H. (1737). *Cahier de mathématique a l'usage de messieurs les officiers de l'école royale de l'artillerie de Strasbourg.* Strasbourg, France: Jean-Renaud Doulssecker.

Hodder, J. (1661). *Hodder's arithmetick, or that necessary art made most easie.* London, UK: R. Davenport.

Hodgson, J. (1723). *A system of the mathematics, containing the Euclidean geometry, plain and spherical geometry, the projection of the sphere, both orthographic and stereographic, astronomy, the use of globes and navigation.* London, UK: Thomas Page.

Hofstadter, R., & Metzger, W. P. (1955). *The development of academic freedom in the United States.* New York, NY: Columbia University Press.

Holbrook, A. (1859). *The normal: Or methods of teaching the common branches, orthoepy, orthography, grammar, geography, arithmetic and elocution, including the outlines, technicalities, explanations, demonstrations, definitions and methods, introductory and peculiar to each branch.* New York, NY: A. S. Barnes & Company.

Hornberger, T. (1968). *Scientific thought in the American colleges 1638–1800.* New York, NY: Octagon Books.

Hoskin, K. (1994). Textbooks and the mathematization of American reality: The role of Charles Davies and the US Military Academy at West Point. *Paradigm, 13*, 11–41.

Houser, M. L. (1943). *Young Abraham Lincoln mathematician.* Peoria, IL: Lester O. Shriver.

Howson, G. (1982/2008). *A history of mathematics education in England.* Cambridge, UK: Cambridge University Press.

Howson, G., Keitel, C., & Kilpatrick, J. (1981). *Curriculum development in mathematics.* Cambridge, UK: Cambridge University Press.

Høyrup, J. (2005a). Review of Maryvonne Spiesser (Ed.), *Une arithmétique commerciale du XVe siècle. Le compendy de la pratique de Barthélemy de Romans. Nuncius, 20*, 481–482.

Høyrup, J. (2005b). Leonardo Fibonacci and Abacco culture: A proposal to invert the roles. *Revue d'Histoire des Mathématiques, 11*, 23–56.

Høyrup, J. (2007). *Jacopo da Firenze's Tractatus Algorismi and early Italian abbacus culture.* London, UK: Birkhäuser Basel/Springer.

Høyrup, J. (2008). The tortuous ways toward a new understanding of algebra in the Italian *Abbacus* School (14th–16th centuries). In O. Figueras, J. L Cortina, A. Alatorre, T. Rojano, & S. Sepulveda (Eds.), *Proceedings of the joint meeting of PME 32 and PME-NA XXX* (Vol. 1, pp. 1–20). Morelia, Mexico: International Group for the Psychology of Mathematics Education.

Hummel, W. W. (1964). Abraham Reincke Beck: Portrait of a schoolmaster. *Journal of the Lancaster County Historical Society, 68*(1), 1–40.

Hutton, C. (1766). *The schoolmaster's guide: Or a complete system of practical arithmetic, adapted to the use of schools, to which is added, a promiscuous collections of questions, and a course of retail book-keeping* (2nd ed.). Newcastle-upon-Tyne, UK: J. White & T. Saint.

Hutton, C. (1812). *A course of mathematics. . ..* New York, NY: Samuel Campbell.

Ifrah, G. (2000). *The universal history of numbers from prehistory to the invention of the computer.* New York, NY: Wiley.

Inglis, A. J. (1911). *The rise of the high school in Massachusetts.* New York, NY: Teachers College.

Isoda, M., (2007). A brief history of mathematics lesson study in Japan. In M. Isoda, M. Stephens, Y. Ohara, & T. Miyakawa (Eds.), *Japanese lesson study in mathematics: Its impact, diversity and potential for educational improvement* (pp. 8–15). Singapore: World Scientific Publishing.

Jackson, L. L. (1906). *The educational significance of sixteenth century arithmetic from the point of view of the present time.* New York, NY: Columbia Teachers College.

Jenkins, J. (1813). *The art of writing*. Cambridge, UK: Author.
Jess, Z. (1799). *A compendious system of practical surveying and dividing of land: Concisely defined, methodologically arranged, and fully exemplified*. Wilmington, DE: Bonsal and Niles.
Jess, Z. (1799/1811). *The American Tutor's Assistant, improved: Or, a compendious system of decimal, practical arithmetic comprising the usual methods of calculation, with the addition of Federal money, and other decimals, dispersed through the several rules of that useful science adapted for the easy and regular instruction of youth in the United States*. Wilmington, DE: Peter Brynberg.
Johnson, C. (1907). *The country school*. New York, NY: Thomas Crowell.
Johnson, H. (1719). *New treatise of practical arithmetic* (4th ed.). London, UK: Thomas Wood.
Jonassen, D. H., & Rohrer-Murphy, L. (1999). Activity theory as a framework for designing constructivist learning environments. *Educational Technology Research and Development, 47*, 61–79.
Jones, H. S. (1954). Foreword by the Astronomer Royal. In E. G. R. Taylor (Ed.), *The mathematical practitioners of Tudor & Stuart England 1485–1714* (pp. ix–x). Cambridge, UK: Cambridge University Press.
Jones, P. S., & Coxford, A. F. (1970). From discovery to an awakened concern for pedagogy, 1492–1821. In P. S. Jones & A. F. Coxford (Eds.), *A history of mathematics education in the United States and Canada* (pp. 11–23). Washington, DC: National Council of Teachers of mathematics.
Kaestle, C. F. (1983/2001). *Pillars of the Republic: Common schools and American society, 1770–1860*. New York, NY: Hill and Wang.
Kamens, D. H., & Benavot, A. (1991). Elite knowledge for the masses: The origins and spread of mathematics and science in national curricula. *American Journal of Education, 99*(2), 137–180.
Karpinski, L. C. (1911). An Italian algebra of the fifteenth century. *Bibliotheca Mathematica, 3*(11), 209–219.
Karpinski, L. C. (1925). *The history of arithmetic*. Chicago, IL: Rand McNally & Company.
Karpinksi, L. C. (1926, February 13). Early arithmetics published in America: A concise history of the rise of the art of figuring things out. *The Dearborn Independent, 26*, 20–24.
Karpinski, L. C. (1929). The Italian arithmetic and algebra of Master Jacob of Florence. *Archeion, 11*, 170–177.
Karpinski, L. C. (1980). *Bibliography of mathematical works printed in America through 1850* (2nd ed.). Ann Arbor, MI: University of Michigan Press.
Katz, M. B. (1968). *The irony of early school reform: Educational innovation in mid-19th century Massachusetts*. Cambridge, MA: Harvard University Press.
Kennedy, M. F., & Harlow, A. F. (1940). *Schoolmaster of yesterday: A three-generation story*. New York, NY: McGraw-Hill.
Kidwell, A. K., Ackerberg-Hastings, A., & Roberts, D. L. (2008). *Tools of American mathematics teaching, 1800–2000*. Washington, DC: Smithsonian Institution and Baltimore, MD: The Johns Hopkins University Press.
Kiely, E. R. (1947). *Surveying instruments: Their history and classroom use (19th yearbook)*. New York, NY: Teachers College Columbia University/National Council of Teachers of Mathematics.
Kilpatrick, J. (1992). A history of research in mathematics education. In D. A. Grouws (Ed.), *Handbook of research on mathematics teaching and learning* (pp. 3–38). New York, NY: The Macmillan Publishing Company.
Kilpatrick, J., & Izsák, A. (2008). A history of algebra in the school curriculum. In C. E. Greenes & R. Rubenstein (Eds.), *Algebra and algebraic thinking in school mathematics: Seventieth yearbook* (pp. 3–18). Reston, VA: National Council of Teachers of mathematics.
Kilpatrick, W. H. (1912) *The Dutch schools of New Netherland and colonial New York*. Washington, DC: US Bureau of Education.

Kinne, W. (1831). *A short system of practical arithmetic, compiled from the best authorities, to which is annexed a short plan of book-keeping, the whole designed for the use of schools* (8th ed.). Hallowell, ME: Glazier, Masters & Co.

Knox, V. (1788). *Liberal education: Or a practical treatise on the methods of acquiring useful and polite learning.* London, UK: Charles Dilly.

Kool, M. (1988). What could we learn from Master Christianus van Varenbraken? In C. Hay (Ed.), *Mathematics from manuscript to print 1300–1600* (pp. 147–155). Oxford, UK: Clarendon Press.

Kool, M. (1999). *Die conste vanden getale. Een studie van Nederlandstalige rekenboeken uit de vijftiende en zestiende eeuw, met een glossarium van rekenkundige termen* [*The art of numbers. A study of Dutch arithmetic books of the 15th and 16th century, with a glossary of arithmetical terms*]. Hilversum, The Netherlands: University College.

Kretschmer, K. (1909). *Die Italienischen portolane des Mittelalters: ein Beitrag zur Geschichte der kartographie und nautik.* Berlin, Germany: Georg Olms Verlagsbuchhandlung Hildesheim.

Lancaster, J. (1805). *Improvement in education, as it respects the industrious classes of the community.* London, UK: Darton & Harvey.

Landis, J. (1853). Examination of teachers' certificates. *Pennsylvania School Journal, 2*(1), 11–12.

Latham, J. L. (1955). *Carry on, Mr Bowditch.* Boston, MA: Houghton Mifflin Company.

Laughlin, S. H. (1845). *A diary of public events and notices of my life and family and my private transactions including travels, readings, correspondence, business, anecdotes, miscellaneous memoranda of men, literature, etc., from January 1st to August 1845, and sketch of my life from infancy.* Library Search Room, Family File, Tennessee State Library and Archives. Nashville, TE.

Lave, J., & Wenger, E. (1998). *Communities of practice: Learning, meaning, and identity.* Cambridge, UK: Cambridge University Press.

Lazenby, M. E. (1938). *Lazenby; being such an account as I have been able to collect of the families in the United States bearing the name.* Washington, DC: Author.

Leavitt, D. (1820). *Leavitt's general improved New-England farmer's almanac and agricultural register, for the year of our Lord.* Exeter, NH: Author.

Leavitt, D. (1826a). *Pike's system of arithmetic abridged: Designed to facilitate the study of science of numbers.* Concord, NH: J. B. Moore.

Leavitt, D. (1826b). *A new ciphering book, adapted to Pike's Arithmetic Abridged, containing illustrative notes, a variety of useful mathematical tables.* Concord, NH: J. B. Moore.

Leavitt, D. (1830). *The teacher's assistant, and scholar's mathematical directory, containing … a solution of the most difficult questions in Pike, Walsh, Adams, Daboll and Robinson's arithmetics, and Flint's Surveying.* Concord, NH: Marsh, Capen and Lyon.

Lee, C. (1797). *The American accomptant; Being a plain, practical and systematic compendium of Federal arithmetic; in three parts; Designed for the use of schools, and specially calculated for the commercial meridian of the United States of America.* Lansingburgh, NY: William W. Wands.

Leybourn, W. (1690). *Of arithmetic.* London, UK: Thomas Basset, Benjamin Tooke, Thomas Sawbridge, Awnsham & John Churchill.

L'Huillier, H. (1976). Eléments nouveaux pour la biographie de Nicolas Chuquet. *Revue d'Histoire des Sciences, 1976*, 347–350.

Littlefield, G. E. (1904). *Early schools and school-books of New England.* Boston, MA: The Club of Odd Volumes.

Loeper, J. J. (1974). *Going to school in 1776.* New York, NY: Athenaeum.

Long, P. O. (2009). Introduction: The world of Michael of Rhodes, ancient mariner. In P. O. Long, D. McGee, & A. M. Stahl (Eds.), *The book of Michael of Rhodes: A 15th maritime manuscript* (Vol. 3, pp. 1–33). Cambridge, MA: MIT Press.

Long, P. O., McGee, D., & Stahl, A. M. (2009). *The book of Michael of Rhodes: A 15th maritime manuscript.* Cambridge, MA: MIT Press.

Lortie, D. (1975). *Schoolteacher: A sociological study.* Chicago, IL: University of Chicago Press.

Lounsbury, R. G. (1930). Yankee trade at Newfoundland. *The New England Quarterly, 3*(4), 607–626.
Love, J. (1688). *Geodaesia, or the art of surveying and measuring of land, made easie, showing by plain and practical rules how to survey, protract, cast up, reduce or divide any piece of land whatsoever; With new tables for the ease of the surveyor in reducing the measures of land. Moreover, a more facile and sure way of surveying by the chain than has hitherto been taught. As also, how to lay out new lands in America, or elsewhere: And how to make a perfect map of a river's mouth or harbour, with several other things never yet published in our language.* London, UK: John Taylor.
Lydon, J. G. (1965). Fish and flour for gold: Southern Europe and the colonial America balance of payments. *The Business History Review, 39*(2), 171–183.
Mahan, A. T. (1898). *The influence of sea power upon history 1660–1783*. Boston, MA: Little, Brown, and Company.
Mann, H. (1845/1925). Boston grammar and writing schools. *The Common School Journal*. In O. W. Caldwell & S. A. Courtis (Eds.), *Then and now in education 1845–1923* (pp. 237–272). Yonkers-on-Hudson: New York, NY: World Book Company.
Mann, H., & Chase, P. (1851). *Arithmetic, practically applied, for advanced pupils, and for private reference*. Philadelphia, PA: E. H. Butler & Co.
Mann, M. P. (1937). *Life of Horace Mann*. Washington, DC: National Education Association.
Marshall, C. (1789). *An introduction to arithmetic: Or the teacher of arithmetic's assistant; containing arithmetic of whole numbers, vulgar fractions, decimal fractions, duodecimals, and three sets of exercising questions*. London, UK: George Herdsfield.
Marshall, H. E. (1956). *Grandest of enterprises: Illinois State Normal University 1857–1957*. Normal, IL: Illinois State University.
Martin, G. H. (1897). *The evolution of the Massachusetts public school system: A historical sketch*. New York, NY: D. Appleton & Co.
Massachusetts Board of Education. (1855). Eighteenth Annual Report of the Secretary. *The New Englander, 13*(1), 43–61.
Massey, W. (1763). *The origin and progress of letters*. London, UK: Johnson.
McCusker, J. J., & Menard, R. R. (1985). *The economy of British America, 1607–1789*. Chapel Hill, NC: University of North Carolina Press.
McTighe, S. (1998). Abraham Bosse and the language of artisans: Genre and perspective in the Académie royale de peinture et de sculpture, 1648–1670. *Oxford Art Journal, 21*(1), 1–26.
Mehl, L. (1968). Die anfange des navigationsunterrichts unler besonderer beriicksichtigung der deutschen verhaltnisse. *Paedagogica Historica International Journal of the History of Education, 2*, 372–441.
Mellis, J. (1582/1658). The third part, or addition to this booke, entreateth of brief rules, called rules of practice, pleasant and commodious effects abridged into a briefer method than hitherto hath been published. In R. Record (Ed.), *Record's arithmetick: Or the ground of arts* ... (pp. 345–535). London, UK: I. Harrison, and H. Bynneman. [Note that Robert Recorde's family name is spelt "Record" and not "Recorde" throughout this publication.]
Meriwether, C. (1907). *Our colonial curriculum 1607–1776*. Washington, DC: Capital Publishing Co.
Meyer, A. E. (1965). *An educational history of the Western world*. New York, NY: McGraw-Hill Book Company.
Michalowicz, K. D., & Howard, A. C. (2003). Pedagogy in text: An analysis of mathematics texts from the 19th century. In G. M. A. Stanic & J. Kilpatrick (Eds.), *A history of school mathematics* (Vol. 1, pp. 77–109). Reston, VA: NCTM.
Middlekauff, R. (1963). *Ancients and axioms*. New Haven, CT: Yale University Press.
Monaghan, E. J. (2001). Literacy instruction and gender in colonial New England. In D. Finkelstein & A. McCleery (Eds.), *The book history reader 1743–1760* (pp. 297–315). London, UK: Taylor & Francis.

References

Monaghan, E. J. (2007). *Learning to read and write in colonial America*. Amhurst, MA: University of Massachusetts Press.

Money, J. (1993). Teaching in the market place, or "*Caesar adsum jam forte: Pompey aderat*": The retaining of knowledge in provincial England during the 18th century. In J. Brewer & R. Porter (Eds.), *Consumption and the world of goods* (pp. 335–377). London, UK: Routledge.

Monroe, P. (1922). *A textbook in the history of education*. New York, NY: The Macmillan Company.

Monroe, W. (Walter) S. (1912). A chapter in the development of arithmetic teaching in the United States. *The Elementary School Teacher, 13*(1), 17–24.

Monroe, W. (Walter) S. (1917). *Development of arithmetic as a school subject*. Washington, DC: Government Printing Office.

Monroe, W. (Will) S. (1969). *History of the Pestalozzian movement in the United States*. New York, NY: Arno Press & The New York Times.

Moore, J., & Perkins, P. (1681). *A new systeme of the mathematicks designed for the use of the Royal Foundation of the Mathematical School in Christ's Hospital*. London, UK: Robert Scott.

Moravian Publication Office. (1876). *Historical sketch of the Moravian Seminary for Young Ladies*. Bethlehem, PA: Author.

Morrice, D. (1801). *The young midshipman's instructor; (designed to be a companion to Hamilton Moore's Navigation): With useful hints to parents of sea youth, and to captains and schoolmasters of the Royal Navy*. London, UK: Knight and Compton.

Morris, R. B. (1946/1966). *Government and labor in early America*. New York, NY: Harper & Row.

Munsell, W. W. (1882). *History of Queen's County, with illustrations, portraits and sketches of prominent families and individuals*. New York, NY: Author.

Nichols, G. (1921). *A Salem shipmaster and merchant: The autobiography of George Nichols*. Boston, MA: The Four Seas Company.

Northend, M. H. (1917). *Memories of old Salem: Drawn from letters of a great-grandmother*. New York, NY: Moffatt, Yard and Company.

Okenfuss, M. J. (1973). Technical training in Russia under Peter the Great. *History of Education Quarterly, 13*(4), 325–342.

Orcutt, H. (1859). *Hints to common school teachers, parents and pupils: Or, gleanings from school-life experience*. Rutland, MA: Geo. A. Tuttle & Company.

Oswald, J. C. (1917). *Benjamin Franklin, printer*. Philadelphia, PA: New York, NY: Doubleday, Page and Company.

Overn, O. E. A. (1937). Changes in curriculum in elementary algebra since 1900 as reflected in the requirements and examinations of the College Entrance Examination Board. *Journal of Experimental Education, 5*, 373–468.

Padfield, P. (2005). *Maritime power and the struggle for freedom: Naval campaigns that shaped the modern world*. Woodstock, NY: The Overlook Press.

Page, C. (1825). *Royal Mathematical School: The elements of navigation* (handwritten). London, UK: Christ's Hospital.

Page, D. P. and the Executive Committee of the State Normal School (1846). State normal school. *District School Journal of the State of New York, 7*(1), 1–5.

Page, D. P. (1877). *Theory and practice of teaching: The motives and methods of good school-keeping* (90th ed.). New York, NY: A. S. Barnes & Company.

Paolitto, D. P. (1976). The effect of cross-age tutoring on adolescence: An inquiry into theoretical assumptions. *Review of Educational Research, 46*(2), 215–287.

Papert, S. (1980). *Mindstorms: Children, computers, and powerful ideas*. New York, NY: Basic Books.

Parker, I. (1914). *Dissenting academies in England: Their rise and progress, and their place among the educational systems of the country*. Cambridge, UK: Cambridge University Press.

Parker, S. C. (1912). *The history of modern elementary education*. Boston, MA: Ginn and Company.

Parsons, W. T. (1976). *The Pennsylvania Dutch: A persistent minority*. Boston, MA: Twayne Publishers.
Peabody Essex Museum. (n.d.). *Salem: Maritime Salem in the age of sail*. Salem, MA: Author.
Pearce, E. H. (1901), *Annals of Christ's Hospital*. London, UK: Hugh Reed.
Peirce, C. (1806). *The arts and sciences abridged with a selection of pieces from celebrated modern authors, calculated to improve the manners and refine the taste of youth; Particularly designed and arranged for the use of schools*. Portsmouth, NH: Author.
Pepys, S. (1995). *The diary of Samuel Pepys, 1662*. London, UK: HarperCollins Publishers.
Perry, J. (1788). *Perry's royal standard English dictionary (being the first work of the kind printed in America) intended to fix a standard for the pronunciation of the English language, conformable to the present practice of polite speakers in Greta Britain and the United States*. Worcester, MA: Isaiah Thomas.
Perry, W. S. (1870). *Historical collections relating to the American colonial church. Volume I: Virginia*. Hartford, CT: Church Press Company.
Phillips, P. L. (1896). *Virginia cartography: A bibliographical description*. Washington, DC: Smithsonian Institution.
Pike, N. (1788). *The new complete system of arithmetic, composed for the use of the citizens of the United States*. Newburyport, MA: John Mycall.
Pike, S. (1811). *The teacher's assistant: A system of practical arithmetic ... to abridge the labour of teachers*. Philadelphia, PA: Johsonn (sic.) and Warner.
Pike, S. (1829). *The teachers' assistant: A system of practical arithmetic ... to abridge the labour of teachers*. Philadelphia, PA: McCarty & Davis.
Plimpton, G. A. (1916). The hornbook and its use in America. *Proceedings of the American Antiquarian Society, 26*, 264–272.
Plumley, N. (1976). The Royal Mathematical School within Christ's Hospital. *Vistas in Astronomy, 20*, 51–59.
Public Schools in Salem. (1839). In H. Mann (Ed.), *The Common School Journal for the year 1839* (pp. 133–135). Boston, MA: Marsh, Capen, Lyon, and Webb.
Radford, L. (2003). On the epistemological limits of language: Mathematical knowledge and social practice during the Renaissance. *Educational Studies in Mathematics, 52*, 123–150.
Randall, W. S. (1975/1976). For better or worse, Franklin made us what we are. In D. R. Boldt (Ed.), *Bicentennial journals: The founding city* (pp. 39–51). Radnor, PA: Chilton Book Company.
Rawley, J. A. (1981). *The trans-Atlantic slave trade*. New York, NY: W. W. Norton & Company.
Recorde, R. (1543/1549). *Grounde of the arts: Teaching the worke and practise, of arithmeticke*. London, UK: R. Wolff.
Recorde, R. (1557/1956). The declaration of the profit of arithmetic. In J. R. Newman (Ed.), *The world of mathematics* (Vol. 1, pp. 212–217). New York, NY: Simon and Schuster.
Reisner, E. H. (1930). *The evolution of the common school*. New York, NY: The Macmillan Company.
Reynolds, B. E. (1993). The algorists vs the abacists: An ancient controversy on the use of calculators. *The College Mathematics Journal, 24*(3), 218–223.
Reynolds, G. H. (1818). *The first elements of arithmetic, or the teacher's and scholar's assistant*. London, UK: Longman, Hurst, Rees, Orme and Brown.
Richeson, A. W. (1935). Warren Colburn and his influence on arithmetic in the United States. *National Mathematics Magazine, 10*(3), 73–79.
Rickey, V. F. (1987). Isaac Newton: Man, myth, and mathematics. *College Mathematics Journal, 18*, 362–389.
Rickey, F. V., & Shell-Gellasch, A. (2010). *Mathematics education at West Point: The first hundred years*. Washington, DC: The Mathematical Association of America.
Ring, B. (1993). *Girlhood embroidery: American samplers and pictorial needlework, 1650–1850*. New York, NY: Knopf Publishers.

Roach, J. (1971). *Public examinations in England 1850–1900.* Cambridge, UK: Cambridge University Press.

Robertson, J., & Wales, W. (1796). *The elements of navigation containing the theory and practice, with the necessary tables and compendiums for finding the latitude and longitude as sea, to which is added a treatise on marine fortification* (6th ed.). London, UK: Christ's Hospital.

Rogers, L. (1999). Conflict and compromise: The evolution of the mathematics curriculum in 19th century England. In P. Radelet-de Grave (Ed.), *Histoire et épistémologie dans l'education mathematique – De la maternelle a l'universite* (pp. 309–319). Louvain-la-Neuve, Belgium: Université d'Ete Européenne sur Histoire et Epistemologie dans l'Education Mathématique.

Rogers, P. (1990). Thoughts on power and pedagogy. In L. Burton (Ed.), *Gender and mathematics: An international perspective* (pp. 38–46). London, UK: Cassell Educational.

Rossi, F. (2009). Michael of Rhodes and his manuscript. In. P. O. Long, D. McGee, & A. M. Stahl (Eds.), *The book of Michael of Rhodes: A 15th century maritime manuscript* (Vol. 3, pp. 99–113). Cambridge, MA: MIT Press.

Rotherham, W. (1852). *The algebraical equation and problem papers, proposed in the examinations of St. John's College, Cambridge from the year 1794 to the present time.* Cambridge, UK: Henry Wallis.

Rudolph, F. (1977). *Curriculum: A history of the American undergraduate course of study since 1636.* San Francisco, CA: Jossey-Bass Publishers.

Sanford, (1927). *The history and significance of certain standard problems in algebra.* New York, NY: Teachers College, Columbia University.

Sangster, J. H. (1865). *Elements of algebra designed for the use of Canadian grammar and common schools.* Montréal, Canada: John Lovell.

Sarton, G. (1957). *Six wings: Men of science in the Renaissance.* London, UK: The Bodley Head.

Schmidt, R. (1993). Die professionalisierung der nautischen Fachbildung: die Seefahrtschule in Bremen, 1799–1969. In J. Brockstedt (Ed.), *Seefahrt an deutschen küsten im Wandel, 1815–1914* (pp. 119–138). Neumtinster, Germany: Wachholtz Verlag.

Schoen, H. H. (1938). The making of maps and charts. In The National Council for Social Studies (Eds.), *Ninth yearbook* (pp. 83–98). Cambridge, MA: The National Council for Social Studies.

Schubring, G. (1987). On the methodology of analyzing historical textbooks: Lacroix as textbook author. *For the Learning of Mathematics, 7*(3), 41–51.

Schubring, G. (2005). *Conflicts between generalization, rigor, and intuition: Number concepts underlying the development of analysis in 17th–19th century France and Germany.* New York, NY: Springer.

Seybolt, R. F. (1917). *Apprenticeship and apprenticeship education in colonial New England and New York.* New York, NY: Teachers College Columbia University.

Seybolt, R. F. (1921). The evening schools of colonial New York City. In New York State Department of Education (Ed.), *Annual Report of the State Department of Education* (pp. 630–652). Albany, NY: State Department of Education.

Seybolt, R. F. (1935). *The private schools of colonial Boston.* Cambridge, MA: Harvard University Press.

S. H. M. (1856). Written examinations. *Rhode Island Schoolmaster, 2*, 156.

Shoup, W. J. (1889). *Graded didactics for teachers' normal institutes.* Saint Paul, MN: H. D. Merrill.

Shunk, F. R. (1840). Common schools of Pennsylvania. *United States Commercial and Statistical Register, 2*(16), 253.

Simons, L. G. (1924). *Introduction of algebra into American schools in the 18th century.* Washington, DC: Department of the Interior Bureau of Education.

Simons, L. G. (1931). The influence of French mathematicians at the end of the 18th century upon the teaching of mathematics in American Colleges. *Isis, 15*, 104–123.

Simons, L. G. (1936a). Short stories in colonial geometry. *Osiris, 1*, 584–605.

Simons, L. G. (1936b). *Bibliography of early American textbooks on algebra published in the colonies and the United States through 1850, together with a characterization of the first edition of each work*. New York, NY: Scripta Mathematics, Yeshiva College.
Sinclair, N. (2008). *The history of the geometry curriculum in the United States*. Charlotte, NC: Information Age Publishing.
Slocomb, W. (1831). *The American calculator, or a concise system of practical arithmetic*. Philadelphia, PA: William Davis.
Small, W. H. (1914). *Early New England schools*. Boston, MA: Ginn.
Smith, D. E. (1900). *The teaching of elementary mathematics*. New York, NY: The Macmillan Company.
Smith, D. E. (1908). *Rara arithmetica*. Boston, MA: Ginn and Company.
Smith, D. E. (1911). Arithmetic. In P. Monroe (Ed.), *A cyclopaedia of education* (Vol. 1, pp. 203–209). New York, NY: The Macmillan Company.
Smith, D. E. (1925). *History of mathematics: Volume II, Special topics of elementary mathematics*. Boston, MA: Ginn and Company.
Smith, D. E. (1933). Early American mathematics. *Tôhoku Mathematical Journal, 38*, 227–232.
Smith, D. E., & Ginsburg, J. (1934). *A history of mathematics in America before 1900*. Chicago, IL: The Mathematical Association of America.
Smith, E. B. (2005). "Let them compleately learn": Manuscript clues about early women's educational practices. In *A manuscript miscellany, Summer 2005 Institute*. Manchester, UK: Manchester University Press.
Smith, R. C. (1831). *Practical and mental arithmetic on a new plan, in which mental arithmetic is combined with the use of the slate*. Boston, MA: Richardson, Lord & Holbrook.
Smith, W. (1973). *Theories of education in early America, 1655–1819*. Indianapolis, IA: Bobbs-Merrill.
Spiesser, M. (Ed.). (2004). *Une arithmétique commerciale du XVe siècle. Le Compendy de la praticque des nombres de Barthélemy de Romans*. Turnhout, Belgium: Brepols Publishers.
State of Massachusetts. (1848). *Report No. 12 to the Massachusetts School Board by Horace Mann*. Boston, MA: Author.
State of Massachusetts. (1855). *Eighteenth annual report of the Secretary of the Massachusetts Board of Education*. Boston, MA: Author.
Sterry, C., & Sterry, J. (1795). *A complete exercise book, in arithmetic: Designed for the use of schools in the United States*. Norwich, CT: John Sterry & Co.
Struik, D. J. (1936). *Mathematics in the Netherlands during the first half of the XVIth century*. David Eugene Smith professional collection, Box 62, Butler Library, Columbia University, New York.
Swetz, F. (1987). *Capitalism and arithmetic: The new math of the 15th century*. La Salle, IL: Open Court.
Swetz, F. (1992). Fifteenth and sixteenth century arithmetic texts: What can we learn from them? *Science & Education, 1*, 365–378.
Swetz, F. (2005). Commercial arithmetic. In T. F. Glick, S. J. Livesey, & F. Wallis (Eds.), *Medieval science, technology and medicine* (pp. 133–135). London, UK: Routledge.
Taton, R. (1986). *Enseignement et diffusion des sciences en France au XVIIIe siècle*. Paris, France: Hermann.
Taylor, E. G. R. (1954). *The mathematical practitioners of Tudor & Stuart England 1485–1714*. Cambridge, UK: Cambridge University Press.
Taylor, E. G. R. (1956). *The haven-finding art: A history of navigation from Odysseus to Captain Cook*. London, UK: Hollis & Carter.
Taylor, E. G. R. (1966). *The mathematical practitioners of Hanoverian England 1714–1840*. Cambridge, UK: Cambridge University Press.
Taylor, P. M. (2002). Implementing the standards: Keys to establishing positive professional inertia in preservice mathematics teachers. *School Science and Mathematics, 102*, 137–142.
Thayer, V. T. (1928). *The passing of the recitation*. Boston, MA: D. C. Heath and Company.

References

Thornton, T. P. (1996). *Handwriting in America: A cultural history*. New Haven, CT: Yale University Press.
Thorpe, F. N. (Ed.). (1893). *Benjamin Franklin and the University of Pennsylvania*. Washington, DC: Bureau of Education.
Thut, I. N. (1957). *The story of education: Philosophical and historical foundations*. New York, NY: McGraw-Hill Book Company.
Thwing, C. F. (1928). *The American and the German university: One hundred years of history*. New York, NY: The Macmillan Company.
Tillinghast, N. (1844). *Elements of plane geometry for the use of schools*. Boston, MA: Lewis & Sampson.
Tillinghast, N. (1852). *Prayers for the use of schools*. Boston, MA: Benjamin B. Musset & Co.
Todd, J., Jess, Z., Waring, W., & Paul, J. (1794/1791). *The American tutor's assistant; or a compendious system of practical arithmetic, containing the several rules of that useful science, concisely defined, methodologically arranged, and fully exemplified: The whole particularly adapted to the easy and regular instruction of youth in our American schools; by sundry teachers in and near Philadelphia*. Philadelphia, PA: Zachariah Poulson.
Tolles, F. B. (1956), Philadelphia's first scientist. *Isis, 45*, 20–30.
Trollope, W. (1834). *A history of the royal foundation of Christ's Hospital*. London, UK: Pickering.
Tuer, A. (1896). *A history of the hornbook* (2 Vols.). London, UK: Leadenhall Press.
Turnbull, H. W. (Ed.). (2008). *The correspondence of Isaac Newton 1676–1687* (Vol. II). Cambridge, UK: The Royal Society and Cambridge University Press.
Tyack, D., & Tobin, W. (1994). The "grammar" of schooling: Why has it been so hard to change? *American Educational Research Journal, 31*(3), 453–479.
Unger, F. (1888). *Die methodik der praktischen arithmetick*. Leipsic, Germany: Teubner.
United States Hydrographic Office. (1943). Nathaniel Bowditch (1773–1838). In N. Bowditch (Ed.), *American practical navigator: An epitome of navigation and nautical astronomy* (p. 3). Washington, DC: Author.
University of Michigan. (1967). *Education in early America*. Ann Arbor, MI: Author.
U.S. Department of Commerce, Bureau of the Census. (1975). *Historical statistics of the United States: Colonial times to 1970*. Washington, DC: Government Printing Office.
Van Berkel, K. (1988). A note on Rudolf Snellius and the early history of mathematics in Leiden. In C. Hay (Ed.), *Mathematics from manuscript to print 1300–1600* (pp. 156–161). Oxford, UK: Clarendon Press.
Van Egmond, W. (1976). *The commercial revolution and the beginnings of Western mathematics in Renaissance Florence, 1300–1500*. PhD dissertation, Indiana University.
Van Egmond, W. (1980). *Practical mathematics in the Italian Renaissance: A catalog of Italian abbacus manuscripts and printed books to 1600*. Firenze, Italy: Istituto E Museo di Storia Della Scienza.
Van Egmond, W. (1988). How algebra came to France. In C. Hay (Ed.), *Mathematics from manuscript to print 1300–1600* (pp. 127–144). Oxford, UK: Clarendon Press.
Van Sickle, J. (2011). *A history of trigonometry education in the United States: 1776–1900*. PhD dissertation, Columbia University.
Venema, P. (1730). *Arithmetica of Cyffer-Konst, volgens de Munten Maten en Gewigten, te Nieu-York, gebruykelyk als mede een kort Ontwerp van de Algebra*. New York, NY: Jacob Goelet.
Vermont Historical Society (1937). From Tunbridge, Vermont, to London, England—The journal of James Guild, peddler, tinker, schoolmaster, portrait painter, from 1818 to 1824. *Proceedings of the Vermont Historical Society, 5*(3), 249–314.
Vinovskis, M. A. (1985). *The origins of public high schools: A reexamination of the Beverly High School controversy*. Madison, WI: The University of Wisconsin Press.
Walkingame, F. (1785). *The tutor's assistant being a compendium of arithmetic and a complete question book* (21st ed.). London, UK: J. Scratcherd & I. Whitaker.

Walsh, M. (1801). *A new system of mercantile arithmetic: Adapted to the commerce of the United States, in its domestic and foreign relations; with forms of accounts, and other writings usually occurring in trade*. Newburyport, MA: Edmund M. Blunt.

Walsh, M. (1820). *A new system of mercantile arithmetic: Adapted to the commerce of the United States, in its domestic and foreign relations; with forms of accounts, and other writings usually occurring in trade* (4th ed.). Salem, MA: John D. Cushing.

Ward, J. (1719). *The young mathematician's guide: Being a plain and easie introduction to the mathematicks*. London, UK: Thomas Horne.

Waters, D. W. (1958). *The art of navigation in England in Elizabethan and early stuart times*. New Haven, CT: Yale University Press.

Watras, J. (2002). *The foundations of educational curriculum and diversity: 1565 to the present*. Boston, MA: Allyn and Bacon.

Watson, F. (1902). The curriculum and text-books of English schools in the first half of the 17th century. *Transactions of the Bibliographical Society VI* (Oct 1900–March 1902) (pp. 159–179). London, UK: Blades, East & Blades.

Watson, F. (1913). The training for seamanship. In P. Monroe (Ed.), *A cyclopaedia of education* (Vol. 5, pp. 309–313). New York, NY: The Macmillan Company.

Watson, F., & Kandel, I. L. (1911). Examinations. In P. Monroe (Ed.), *A cyclopaedia of education* (Vol. 2, pp. 532–538). New York, NY: The Macmillan Company.

Wayland, F. (1842). *Thoughts on the present collegiate system*. Boston, MA: Gould, Kendall & Lincoln.

Webb, S. (2008). *Country school copybook*. Birmingham, AL: Author.

Webster, N. (1828). *American dictionary of the English language*. New Haven, CT: S. Comverse.

Webster, W. (1740). *Arithmetick in epitome: Or a compendium of all its rules, both vulgar and decimal* (6th ed.). London, UK: D. Browne.

Wells, H. W. (1900). *The schools and the teachers of early Peoria*. Peoria, IL: Jacquin.

White, Burditt & Co. (1809). *Ciphering book* (unlined, blank pages). Boston, MA: Author.

White, E. (1886). *The elements of pedagogy*. New York, NY: American Book Company.

Whitrow, G. J. (1988). Why did mathematics begin to take off in the sixteenth century? In C. Hay (Ed.), *Mathematics from manuscript to print 1300–1600* (pp. 264–269). Oxford, UK: Clarendon Press.

Wickersham, J. P. (1886). *A history of education in Pennsylvania*. Lancaster, PA: Inquirer Publishing Company.

Williamson, R. S. (1928). Grammar school arithmetic a century ago. *The Mathematical Gazette, 14*(194), 128–133.

Wilson, J. I. (1842). *A brief history of Christ's Hospital, from its foundation by King Edward the Sixth* (7th ed.). London, UK: John Van Voost.

Wingate, E. (1630). *Arithmetique made easie*. London, UK: Stephens and Meredith.

Wingate, E. (1671). *The Clark's tutor for arithmetick and writing, or, a plain and easie way of arithmetick*. London, UK: Stephens and Meredith.

Workman, B. (1789). *The American accountant; or schoolmasters' new assistant*. Philadelphia, PA: John McCulloch.

Wright, P. W. D., Wright, P. D., & Heath, W. H. (2008). *Wrightslaw: No Child Left Behind*. Hartsfield, VA: Harbor House Law Press.

Wroth, L. C. (1952). Some American contributions to the art of navigation. *Proceedings of the Massachusetts Historical Society* (Vol. 68, pp. 72–112). Boston, MA: Massachusetts Historical Society.

Yeldham, F. A. (1936). *The teaching of arithmetic through four hundred years (1535–1935)*. London, UK: Harrap.

Zimmerman, J. (2009). *Small wonder: The little red schoolhouse in history and memory*. New Haven, CT: Yale University Press.

Author Index

A

Ackerberg-Hastings, Amy, 45, 62, 82, 96, 100, 109
Adams, Daniel, 32, 58, 111, 116–117, 122, 126–128, 137, 145, 161, 171, 178, 191
Adams, Oscar Fay, 115
Alexander, Johannes, 66
Allan, George A. T., 26, 66, 80
Allen, John B. L., 30–31, 33, 44, 63, 66
Allison, Robert J., 149
Ames, Glen Joseph, 7
Ames, Susie M., 40
Andrews, Charles C., 121
Andrews, Charles M., 47
Angus, David L., 135
Archer, Léonie J., 8
Arrighi, Gino, 9
Arrowood, Charles F., 41, 69, 76
Ayres, John, 21, 29

B

Babu, D. Senthil, 71, 144
Bache, Alexander Dallas, 25–26, 29, 31
Bailyn, Bernard, 39, 42
Baker, Humfrey, 10, 18, 33, 176
Bangs, Jeremy, 19
Barnard, Henrys, 131, 133–134, 136, 145
Barrême, Nicolas, 83
Beaujouan, Guy, 19
Beckers, Danny J., 144
Bellhouse, David R., 9, 11–12, 96
Benoit, Paul, 14, 19
Bickham, George, 21, 31, 142
Bjarnadóttir, A. Kristín, 19
Boethius, Anicius Manlius Severinus, 18
Bonnycastle, John, 97, 106, 111, 113, 123, 126, 150, 172, 181
Bosse, Abraham, 20
Bowditch, Nathaniel, 45, 47, 61, 65, 67, 70

Bremner, Robert H., 39
Brockliss, Laurence W. B., 82
Brooks, Baylus C., 26, 66
Brooks, Edward, 134, 137, 183
Broome, Edwin C., 77–78, 97
Brown, Joseph Clifton, 9, 12
Burke, Edmund, 7–8, 95
Burke, William, 7–8, 95
Burrowes, T. H., 34, 75–76
Burton, Warren, 111, 117, 124
Burtt, Edward H., 77
Butler, William, 21–23, 31–32, 147

C

Cajori, Florian, 21, 59, 74, 81, 97, 107, 119, 125, 133, 150
Caldwell, Otis W., 129–130, 136
Campbell, Douglas, 19
Carlo, Paula Wheeler, 21, 88
Carpenter, Charles, 20–21, 34, 141
Chambers, Reuben, 123–124, 148
Champion, Richard J., 21, 175
Chase, Pliny Earle, 32, 185
Child, John, 43
Church, Albert, 97
Clarke, James Freeman, 120–121
Clason, Robert Grant, 108
Clemens, Samuel, 1
Clements, M. A. (Ken), 32, 41, 60–64, 75, 79, 107, 126, 142–143, 156
Cobb, Lyman, 123, 148
Cockburn, J. S., 24, 97
Cocker, Edward, 21, 29, 38, 96, 107, 113, 142
Coffman, Lotus Delta, 9, 12
Cohen, Patricia Cline, 17, 37–38, 40, 43, 72–75, 81–82, 93, 95, 97, 107, 117
Colburn, Dana Pond, 132–134
Colburn, Warren, 83, 116, 119–121, 126, 137, 140, 145, 149–150

213

Coldham, Peter Wilson, 26, 41, 66
Collinder, Per Arne, 7
Colson, John, 21
Connor, Ralph, 76–77
Cook, James, 26
Coolidge, Olivia E., 34
Coon, Charles L., 40, 91
Cornell, Ezra, 147–148
Courtis, Stuart A., 129–130, 136
Cowley, Elizabeth B., 9
Coxford, Arthur F., 112
Cracraft, James, 26
Cremin, Lawrence A., 21, 24, 41–42
Cross, Anthony Glen, 26
Cubberley, Ellwood P., 17, 21, 73, 111, 121
Curbera, Guillermo P., 17

D

d'Ambrosio, Ubiritan, 71
Daboll, Nathan, 111, 122–123, 126, 137, 145, 161, 163–164, 168–171, 174, 180
da Firenze, Jacopo, 12
Da Silva, Maria C. L., 82
Dash, Joan, 16
Davenport, Bishop, 74
Davies, Charles, 111, 116, 126, 134, 137, 145, 150, 181
Davis, William E., 10
Day, Jeremiah, 45, 100
De Morgan, Augustus, 11, 42, 74, 148
DeGarmo, Charles, 109
Dewalt, Mark William, 19, 43
Dickens, Charles, 77, 112, 148
Dilworth, Thomas, 33, 96, 150
Ditton, Humfry, 66
Dix, Dorothea, 121
Dixon, T., 24
Doar, Ashley K., 40, 59, 91, 93, 119, 140–141, 150
Dodd, James B., 134–136
Donoghue, Eileen F., 102, 136
Dossey, John, 145
Douglas, Paul Howard, 39
Dowling, Daniel, 33, 146
Durkin, Joseph T., 16, 124
Dyer, Elisha, 111, 126

E

Earle, Alice Morse, 34, 38, 42, 140
Eaton, James, 115
Eby, Frederick, 41, 69, 76
Edwards, Richard, 131–134, 136
Eggleston, Edward, 76–77, 106

Ellerton, Nerida F., 32, 37, 41, 60–64, 75, 79, 107, 126, 142–143, 153, 156
Emerson, Frederick, 83, 106, 123, 126, 129, 137
Emigh, Rebecca Jean, 14, 40, 71

F

Falola, Toyin, 7
Farrar, John, 45
Fayazmanesh, Sasan, 10
Fink, Karl, 14, 23, 33, 43, 52, 133
Finkelstein, Barbara, 133
Fisher, George, 38
Flibbert, Joseph, 61
Flint, Abel, 122
Fowle, William B., 111–112, 121
Fox, Joseph, 121
Franci, Raffaella, 6, 8, 12–14, 81
Freeman, F. N., 21, 29
French, John Homer, 103
Fuess, Claude M., 115

G

Gay, Ebenezer, 45, 114
Gaydos, Tamara, 60–61, 63–64, 152
Gibbs, George, 47
Gibson, Robert, 111
Gies, Frances, 10
Gies, Joseph, 10
Gilman, Caroline, 106, 173, 182
Ginsburg, Jekuthiel, 42–43, 68, 81
Glaeser, Georges, 82
Goodrich, Samuel G., 38, 42, 108, 120, 150
Goss, David, 61
Greenleaf, Benjamin, 111, 134, 137, 145
Greenwood, Isaac, 31, 45, 83, 97, 116
Grendler, Paul F., 14
Griffis, William Elliot, 43
Griffith, Patrick, 31, 134
Grove, Myrna J., 125
Gummere, John, 111
Guralnick, Stanley M., 74

H

Hale, Edward Everett, 121
Halvorsen, Katherine T., 145
Halwas, Robin, 74, 93
Hans, Nicholas A., 24–26, 30
Harlow, Alvin F., 108
Harper, Charles A., 131, 133–137
Hay, Cynthia, 23
Heal, Ambrose, 21, 33
Heath, Suzanne W., 145

Heeffer, Albrecht, 12, 102, 142
Hendrick, Elijah H., 60
Henry, James, 127, 131, 175
Herttenstein, J. H., 31
Hill, Thomas P., 72–73
Hodder, James, 21, 29, 38, 96, 107, 142
Hodgson, James, 25–26, 28, 66–67
Hofstadter, Richard, 74–75
Holbrook, Alfred, 124, 133–134
Hornberger, Theodore, 76, 100
Hoskin, Keith, 117, 135
Houser, M. L., 34, 111–112
Howard, A. C., 95, 108, 110, 134, 136
Howson, A. Geoffrey, 6, 11, 17–20, 23–26, 29–31, 41, 80, 96, 107, 115
Høyrup, Jens, 4, 9, 12–13
Hummel, William W., 135
Hutton, Charles, 31–32, 97, 142, 166

I

Ifrah, Georges, 11–12, 18–19, 24
Inglis, Alexander J., 136
Isoda, Masami, 146
Izsak, Andrew, 100

J

Jackson, Lambert Lincoln, 10, 18
Jänicke, E., 112
Jenkins, John, 21
Jess, Zachariah, 59, 96, 126, 137, 145, 166, 174–176
Johnson, Clifton, 77
Johnson, Humphry, 21, 31
Jonassen, David H., 71
Jones, Harold Spencer, 17, 24–25
Jones, Phillip S., 112

K

Kaestle, Carl F., 42
Kamens, David, 40
Kampas, Barbara, 60–61, 63–64, 152
Kandel, Isaac Leon, 76, 127
Karpinski, Louis C., 9, 20, 24, 29–30, 38, 40, 44, 59, 90, 79, 81, 87, 98, 101, 107–108, 111, 113, 120, 126–127, 136, 150
Katz, Michael B., 129–131, 133, 135–136
Keitel, Christine, 7
Kennedy, Millard F., 108
Kidwell, Peggy A., 109–110, 133
Kiely, Edmond R., 101
Kilpatrick, Jeremy, 7, 100, 129, 131
Kilpatrick, William H., 43
King, H. P. F., 24

Kinne, William, 116
Knox, Vicesimus, 18, 76
Kool, Marjolein, 19
Kretschmer, Konrad, 7–8

L

L'Huillier, H., 14
Lancaster, Joseph, 75, 121, 163, 166
Landis, J., 131
Latham, Jean Lee, 61–62, 65, 70–71
Laughlin, Samuel Hervey, 43, 106, 113
Lave, Jean, 71
Lazenby, Mary Elinor, 111, 150–151
Lazenby, Robert, 150, 151
Leavitt, Dudley, 48, 122–123
Lee, Chauncey, 83
Leybourn, William, 38, 96, 108
Lincoln, Abraham, 34, 111
Littlefield, George Emory, 20, 34, 107, 148
Loeper, John J., 39
Long, Pamela O., 7–9, 11, 14, 49, 81
Lortie, Dan, 1
Lounsbury, Ralph G., 52
Love, John, 111, 165
Lydon, James, 52

M

Mahan, Alfred Thayer, 7, 17, 26, 30
Malcolm, Alexander, 42, 134
Mann, Horace, 29, 32, 119, 121, 123, 127, 129–131, 134, 136, 145, 148
Marshall, Charles, 22, 49, 124
Marshall, Helen E., 134
Martin, George H., 119
McAlister, Jim, 61
McCrone, Sharon S., 145
McCusker, John J., 52
McDonnell, K. G. T., 24
McGee, David, 7
McTighe, Sheila, 20
Mehl, Lothar, 16
Mellis, John, 106
Menard, Russell R., 52
Meriwether, Colyer, 3, 68, 74, 84, 107, 111, 127
Metzger, Walter P., 74–75
Meyer, Adolph E., 43–44, 63, 76
Michalowicz, Karen D., 95, 108, 136
Middlekauff, Robert, 60, 74, 93, 107, 112
Mirel, Jeffrey E., 135
Money, J., 19
Monroe, Paul, 23

Monroe, Walter S., 17, 21, 37, 43, 73–74, 81, 98, 107–108, 110–111
Monroe, Will Seymour, 131
Moore, Jonas, 30, 66
Morrice, David, 17, 91
Morris, Richard Brandon, 43
Munsell, William W., 107, 112

N

Newton, Isaac, 23–24, 26, 30–31, 41–42, 80
Nichols, George, 45, 61, 69, 77
Northend, Mary Harrod, 34

O

Okenfuss, Max J., 26
Orcutt, Hiram, 136
Oswald, John Clyde, 3, 45–46
Overn, O. E. A., 100

P

Padfield, Peter, 26
Page, Charles, 25, 28, 41, 66–67, 112, 128, 131–134, 136, 148
Page, David P., 112, 131–134, 148
Paolitto, Diana P., 121
Papert, Seymour, 1
Parker, Irene, 24, 111
Parker, Samuel C., 24, 38–39
Parsons, William T., 43, 87
Patoun, Archibald, 65
Paul, Jeremiah, 14, 59
Pearson, Eliphalet, 114
Peirce, Charles, 69
Pepys, Samuel, 29–30, 80
Perry, William, 38
Pestalozzi, Johann Heinrich, 109, 119–120, 131–132, 146
Phillips, Philip Lee, 44
Pike, Nicolas, 33, 58, 83, 96, 111, 145, 150
Pike, Stephen, 32, 111, 126, 137, 145
Plimpton, George A., 24, 42, 140
Plumley, N. P., 16, 25
Prust, Thomas, 52–53, 102, 104, 155, 157

Q

Quincy, Josiah, 115

R

Radford, Luis, 8–9
Randall, Willard S., 71
Rawley, James A., 46–47, 61
Ray, Joseph, 111, 126, 137, 145

Record(e), Robert, 18, 23–24, 33, 106–108
Reinhard, Andréas, 13
Reisner, Edward H., 131, 133–134, 136
Reynolds, Barbara E., 18
Reynolds, George H., 18, 21, 29, 142
Richeson, A. W., 133–134, 136, 149
Rickey, V. Frederick, 80, 97
Rigatelli, Laura Toti, 8, 13
Ring, Betty, 77
Roberts, David Lindsay, 109
Roberts, William, 68
Robertson, John, 28
Rogers, Leo, 96
Rogers, Pat, 43
Rohrer-Murphy, Lucia, 71
Rossi, Franco, 103
Rotherham, W., 127
Rowlandson, Thomas, 71–72
Rudolph, Frederick, 91, 100

S

Sanford, Vera, 12
Sangster, John Herbert, 134
Sarton, George, 12, 19
Schmidt, Rüdiger, 16
Schoen, Harriet H., 41
Schubring, Gert, 31, 110
Seybolt, Robert, 3, 38–40
Shell-Gellasch, Amy, 97
Shoup, William J., 135
Shunk, Francis Raun, 35
Simons, Lao Genevra, 21, 63, 68, 74, 81–82, 97, 100
Sinclair, Nathalie, 35, 63, 65, 68–69
Slocomb, William, 111, 113
Small, W. H., 21, 74, 126
Smith, David Eugene, 9, 12–13, 18, 23, 42–45, 68, 81, 83, 109–110, 125, 134, 150
Smith, Emily Bowles, 21, 48, 83, 93
Smith, Roswell C., 106, 116, 126
Smith, Wilson, 134, 136
Spiesser, Maryvonne, 11–12
Stahl, Alan M., 7
Sterry, Consider, 32, 116
Sterry, John, 32, 116
Stoddard, John, 137
Struik, Dirk, 20
Swetz, Frank, 7–9, 11, 13, 18–19, 59, 80

T

Taton, Rene, 31
Taylor, P. Mark, 1
Thayer, Vivian Trow, 134, 145, 147

Author Index

Thornton, Tamara P., 29, 39, 72
Thorpe, Francis N., 71
Thut, Isaak Noah, 23
Thwing, Charles Frank, 44
Tillinghast, Nicholas, 131–134
Tobin, William, 125, 135
Todd, John, 59
Tolles, Bryant F., 61, 108
Trask, Richard B., 61
Trollope, William, 26–27, 29–30, 33, 67, 76, 80
Tuer, Andrew W., 42
Turnbull, Herbert W., 26, 41, 102–103
Tyack, David, 125, 135
Taylor, Eva Germaine, 7–8, 16–17, 19, 24, 30, 41

U

Unger, Friedrich, 11, 32, 37

V

Valente, Wagner R., 82
Van Berkel, Klaas, 19, 25
Van Egmond, Warren, 6, 8, 11–16, 23, 32–33, 59, 81, 102, 141–143, 152
Van Sickle, Jenna, 69, 74
Venema, Pieter, 20–21, 38
Vinovskis, Mans A., 135–136

W

Wales, William, 26, 28
Walkingame, Francis, 96, 106, 111–112, 126

Walsh, Michael, 60, 62, 83, 122, 126, 137, 145
Ward, John, 97
Waring, William, 59
Warnock, Amanda, 7
Washington, George, 34, 148, 167
Waters, D. W., 17, 25–26, 65
Watras, Joseph, 135
Watson, Foster, 21, 24, 30, 76, 127
Wayland, Francis, 74, 131
Webb, Susan, 109
Webster, Noah, 5, 106, 126
Webster, William, 21
Wells, Hubert Wetmore, 108
Wendell, Abraham, 97
Wenger, Etienne, 71
White, Burditt & Co., 71
White, Emerson E., 71, 131, 134–135, 145
Whitrow, Gerald J., 23, 59
Wickersham, James P., 119–112, 125
Williamson, R. S., 5, 74, 107
Wilson, John Iliff, 26–27, 72, 91, 140, 150
Wingate, Edmund, 29, 32, 38, 96, 107
Workman, Benjamin, 112
Wright, Pamela, 145
Wright, Peter W. D., 145
Wroth, Lawrence C., 44

Y

Yeldham, Florence A., 5, 18, 74, 106–107, 146

Z

Zimmerman, Jonathan, 108, 177

Subject Index

A
Abacus board, 6
Abbaci and the *abbaco* tradition, 10–16, 20–21, 24, 29, 32–33, 59–60, 62, 81, 84, 96–97, 102, 106, 119, 139–142, 144, 148
Africa, 8, 10, 15, 45–47, 144, 153
Algebra, 4–5, 9–10, 12–15, 21, 27–28, 30, 42, 54, 66, 69, 80, 93–94, 96, 100–102, 113, 115, 120, 131, 141–142, 146, 158, 162, 166, 174, 177–178, 182
Algorism, 18–19
Algorithm, 10, 12, 18, 23, 53–54, 83–84, 141, 143–144, 162, 166–167, 170, 179
Alligation, 8, 14–15, 17, 47, 54–55, 59, 62, 70–71, 80–81, 98, 141, 151, 155–157, 160–161, 164, 168–169, 172–173, 177, 181, 190
Almanac, 18, 48, 91, 122, 151
Amsterdam, 8, 43–44
Applied mathematics, 5, 17, 31, 44
Apprentice, 2–3, 9, 16, 26, 31, 39–46, 63–64, 70–72, 87, 91, 102, 114, 125, 140, 143–144, 151–152
Arabic influence on arithmetic, 4, 8, 10, 14–15, 18–19, 59, 141, 144
Arithmetic, 9, 21–22, 28, 42, 46–47, 59–62, 75, 78, 96, 107, 113, 116–117, 119, 122, 126, 130, 132, 137, 155–183, 188, 191
Arkansas, 89, 95, 182
Attendance at school, 34, 125
Attitudes
  to mathematics education, 19–24, 42
  to mathematics, 17–18

B
Baltimore, 57, 177–179
Bethlehem Digital History Projects, 93

Blackboard, 108, 124, 133, 135, 150
Bridgewater Normal School, 131
Bureau of Education, 100
Butler Library (Columbia University), 24, 31, 72–73, 82, 91
Business arithmetic, 3, 9–10, 81, 98, 114, 161
Bookkeeping, 8, 20, 42, 69, 115
Books, *see* Textbooks
  Latin School, 71, 88, 120
  Schoolmasters (1840s), 17, 129, 131, 134–136
Boston, 17, 45–47, 57, 59, 62, 64, 67, 71, 73, 88, 95, 108, 111, 120–121, 124, 129–131, 134–136, 147–149, 151, 160, 166, 170, 175
Barter, 10, 14–15, 17, 46–47, 55, 59, 62, 70, 98, 122, 141, 147, 155, 170, 175–180, 190
Brokerage, 17, 59, 190

C
Calligraphy, 16, 21, 24, 26, 28–31, 39, 49–50, 62, 64, 67, 72, 77, 80, 101, 127, 141–142, 148, 156, 160, 169, 171, 176
Caribbean, 8, 41
China, 61, 65
Christ's Hospital, London, 24–31, 33, 37, 41, 43–44, 63–64, 66–68, 70, 73, 76, 80–81, 102, 142, 161
  and Charles Page, 25, 27–28, 41, 67, 80
  curriculum of, 25, 28–30, 63, 66–68, 80–81
  cyphering books from, 24–31, 37, 41, 43–44, 63–64, 66–68, 70, 73, 76, 80–81, 142
  evaluation of students at, 29, 127–131, 145
  influence on American, 31, 107
    education of, 2, 19–23, 44, 66, 146
  navigation emphasis at, 68

219

Christ's Hospital, London (*cont.*)
    writing emphasis at, 18, 29
    See also Royal Mathematical School
Chronology, 28, 64
Ciphering, *see* Cyphering
Cipher book, *see* Cyphering book
Classical tradition in curriculum, 10
Clements Library (University of Michigan), 31, 41, 72–73, 82, 91, 140–141
College of William and Mary, 45, 72, 82, 91, 106, 140
Colonial period, 3
Columbia University (King's College), 24, 31, 72–73, 82, 85, 91, 97
Compound operations, 17, 59, 71, 91, 98, 155, 168–170, 174, 176, 181–182
Connecticut, 49, 60, 89, 108, 124, 133, 164
Cooper, 61, 70, 161
Copybook, *see* Cyphering Book
Currency conversion, 10, 98, 141, 147, 160
Curriculum
    implemented, 83–84, 126, 151
    intended, 35, 84, 126
Cyphering approach
    advantages of, 8, 84, 113–114, 139, 146, 148
    in Continental Europe, 34, 95
    criticisms of, 133
    demise of, 119–137, 145–146, 149, 152
    model for, 11, 31–35
    in North America, 2–6, 31, 35, 37–78, 81, 87–88, 92–93, 110–111, 127, 134, 140, 142, 146, 151–152, 155
Cyphering books
    at Christ's Hospital, 24–31, 41, 43–44, 63, 66–68, 70, 73, 76, 80–81, 102, 142
    Chichester-Pine manuscript, 49–50, 52
    complemented by textbooks, 30–31, 126, 129
    Ellerton/Clements principal data set, 107
    evaluative function of, 76–78, 134
    genres within, 64, 70, 156, 176, 178, 180, 183
    Michael of Rhodes, 12, 14, 81
    Navigation, 5, 7, 16, 30, 41–42, 44–45, 47, 60–61, 63–71, 76, 80, 96, 100–102, 113, 141, 161, 176, 179–180
    often prepared in winter, 38, 106, 150, 157, 159–165, 167–172, 174–179, 181–183
    terminology, 57
    Thomas Prust manuscript, 52–53, 102, 157

Cyphering tradition
    as curriculum "glue", 151
    cyphering schools (Russia), 26
    related to *abbaco* tradition, 14, 21, 24, 29, 32, 59–60, 62, 81, 84, 96–97, 119, 139–140, 152

D

Dame Schools, 2, 17, 42, 125, 151
Decimals, 12, 17, 33, 59, 80, 91, 98, 101, 106–107, 130, 141, 155–156, 160, 167, 170, 173, 181–182, 185, 188–190
Diagrams, 15–16, 64, 77, 103, 141, 157–183
Discipline in schools, 121
Dissenting academies (Great Britain), 24, 96
Duodecimals, 59, 80, 98, 156, 170
Dutch East India Company, 7
Dutch education traditions, 7–9, 19

E

East Indies, 16–17, 26, 47, 65, 67, 71
English Department (Phillips Exeter), 115
Entrance requirements of colleges, 76
Equations (algebra), 174, 177, 182
Equation of payments, 15, 54–55, 59, 62, 80, 98, 135, 141, 147, 155, 170, 179, 189–190
Ethnomathematics, 2, 40, 70–73
Euclidean geometry, 28, 61, 69, 151
Evaluation
    at Christ's Hospital, 29
    by committee, 80–81, 127
    role of ciphering books, 5, 79, 108, 113, 122
Evening classes, 34, 39, 87, 91, 102, 112
Exercise books, 5–6, 79, 127, 142, 174, 179

F

Fellowship, 8, 14–15, 49–50, 55, 59, 62, 70, 80–81, 98, 115, 141, 151, 155–156, 160, 187, 190
Four operations, 8, 14, 35, 46, 48, 53, 55, 75, 91, 97, 100, 119, 141, 147, 155
    in algebra, 14, 100, 141
Fractions, 8, 12, 15, 17–18, 21, 32–33, 59, 77–78, 91, 98, 100, 107, 130, 141, 147, 155, 162–164, 167, 172–173, 176–177, 179–183, 185–186, 188–190
    in algebra, 21, 100
France, 7, 10–11, 19, 26, 31, 44, 48, 67, 83, 87

Subject Index

G

Gauging, 8, 58–59, 62, 64–65, 70, 80–81, 98, 151, 155, 169
Gender and mathematics, 93, 155
Genre
    IRCEE, 70, 113, 122, 151, 183
    PCA, 49, 52–53, 64, 70, 165, 176, 178–180, 183
    Rhetorical, 11, 141
Geometry, 4–5, 8–10, 24, 26–28, 30–31, 42, 47–48, 60–61, 63, 65, 67–69, 80, 96, 100–102, 113, 115, 132, 141, 146, 151, 159, 161, 165–166, 170, 176, 178–180, 187, 189–190
Germany, 8, 43
Girls and arithmetic, 43, 48, 53, 93, 146
Grade(d) schools, 92, 125, 133, 135, 145
Great Britain (background), 5, 7, 25–26, 30–31, 48, 61, 66–68, 95–96, 107, 109, 140, 142, 146, 152

H

Hanseatic League, 8–9
Harvard College (University), 17, 31, 41, 44–45, 67–69, 77–78, 115, 164
Hindu-Arabic numerals, 4, 10, 15, 18, 81, 144
Holland, 10, 19–20, 25, 43
Hornbook, 20, 42, 44, 107, 140, 151
Houghton Library (Harvard), 17, 31, 67–68, 72, 82, 91–93, 111, 117, 140–141
Huguenots and arithmetic, 20, 91, 140

I

Iceland, 19
Illinois State Normal University, 131, 134
India, 2, 7, 17, 43, 61, 144, 153
Indiana, 89, 95, 108, 111, 165
Intellectual (mental) arithmetic, 106, 132
Interest, simple and compound, 8, 15, 49, 141, 147, 155
Involution and evolution, 59, 141, 156, 167
IRCEE genre, 70, 113, 122, 151, 183
Italy, 4, 9, 11–13, 19

J

Japan, 146
Journals, 28, 30, 38, 65, 79, 124, 131–132, 152–153, 176, 180

L

Latin, 9–10, 17, 20, 24–25, 44, 66, 71, 88, 91, 102, 114–115, 120
Leiden, 8, 19, 25

Leonardo Pisarno ("Fibonacci"), 144
Liber Abbaci, 10, 59, 144
Logarithms, 12, 16, 24, 63–65, 69, 98, 100–101, 141, 161, 172, 174, 187
Logs of journeys, 65
London, 8, 10–11, 17–18, 22–29, 31, 41–42, 44, 53, 63, 65, 67, 70–71, 73, 76, 80, 82, 104, 108, 110, 113, 166, 192–193
Longitude, 16, 28, 44
Loss and gain, 14–15, 17, 49, 59, 62, 98, 141, 147, 155, 164, 175, 177–178, 190

M

Madeira, 28, 67–68
Maryland, 44, 52, 59, 89, 95, 139, 150, 162–163, 166, 170, 172, 175–176
Massachusetts, 19, 31, 43, 45, 60–62, 77, 82, 85, 88–89, 91, 107–108, 114, 121, 127, 129, 131, 134–137, 152, 165–166, 170, 172, 183
Massachusetts Board, 127, 129, 136–137
    of education, 127, 129, 137
Mathematical activity, 2, 5–6
Mathematics education, 1–2, 5–9, 11–14, 18–24, 29–31, 34–35, 46, 61, 70–71, 73, 78–79, 81–82, 97, 102, 110, 113–116, 132, 134, 136–137, 144, 147, 150–154
Measurement (elementary), 114
Mercantile arithmetic, 14, 40, 60, 62, 146
Michael of Rhodes, 12, 14, 81, 103
Money, 9, 15, 49, 53, 98, 174, 187–189, 191
Monitorial system, 121
Moravian Publishing office, 93

N

Naval schoolmasters, 16–17
Navigation cyphering books, 30, 60–61, 63–71
Navigation education, 60, 63–66
New England, 1, 19, 38, 40, 44, 46, 48, 60–61, 63–64, 69, 71, 76, 83, 91, 93, 95, 100, 106, 136, 139, 150
New Hampshire, 61, 89, 122, 164, 166, 178
New York, 3, 8, 20–21, 31, 40, 42–46, 54, 57, 71–72, 78, 89, 91, 95, 98, 106, 110, 113, 130, 132, 134–135, 139–140, 147, 150–151, 157, 161, 165, 168, 170, 172–173, 175, 178, 189
Normal schools, 92, 100, 113, 119, 124, 131–137, 145–146, 149
    opposition to the ciphering tradition, 131, 136

Normal schools (*cont.*)
  emphasis on whole-class instruction, 39, 112, 133, 150
"No Child Left Behind", 145
North Carolina, 72, 82, 85, 89, 91, 93, 95, 106, 124, 140–141, 148, 150–151, 171, 179
Notation and numeration, 97, 141
Notebooks, 5, 17, 68, 79, 82, 97, 152

O
Ohio, 89, 95, 134–135, 163, 169, 172–174, 177

P
Paper, 4, 14, 17, 38, 75, 97, 107–109, 112, 123–124, 127, 130, 161–164, 166, 168, 170–173, 177, 179
PCA genre, 49, 52–53, 64, 70, 165, 176, 178–180, 183
Peabody, 31, 47, 60–61, 141, 152
Penmanship, 30, 39, 49–50, 62, 64, 67, 72, 77–78, 80, 101, 107, 114, 123, 127, 141–143, 148, 156–183
Pennsylvania, 34–35, 40, 43–44, 59, 75–76, 87, 89, 93, 95, 106, 108–109, 123, 125, 131, 137, 139, 150, 161, 166–167, 170–171, 177–181
  University of, 137
Percentage, 59, 98, 130, 140, 155
Permutations and combinations, 59, 98, 141, 156
Philadelphia, 45–46, 57, 59, 68, 71, 95, 135, 137, 160, 164, 167–168, 171, 175–177, 180–181
Phillips Academy (Andover, MA), 114–115
Phillips Library (Salem, MA), 16, 31, 41, 60–67, 71–73, 82, 85, 91–93, 140–141, 152
Pinnock's *Catechisms*, 113
Population statistics, 45
Position, single and double, 57, 59, 98, 155–156
Practical mathematicians, 7, 17
Practice, 1–2, 7, 9–12, 18, 21, 28, 37, 40, 46, 49, 52–53, 63, 67, 71, 74–76, 80–82, 91–92, 97–98, 106, 110, 112, 115–116, 124, 141, 146, 148, 150–151, 155–156, 159, 162, 166, 168, 170, 187–188
Princeton University, 78
Principal Data Set (PDS), 2, 4–5, 39, 48, 83, 87–117, 152, 155–183
Progressions (arithmetical and geometrical), 47, 59, 98, 100, 141, 188
Promiscuous questions, 143, 188
Public Schools in Salem, 2–3, 9, 40, 43, 62, 71, 91, 137

Q
Quaker, 45

R
Recitation, 20, 74, 97, 113–117, 121–122, 133, 135–136, 145, 148, 150
Reckoning masters, 6, 9–10, 15–16, 18, 32, 142
Recreational mathematics, 10, 15, 100, 176
Reduction, 17, 50–51, 55, 59, 71, 75, 78, 80, 91, 97–98, 141–142, 155, 163–165, 167–168, 170–173, 175, 177–179, 181–182, 186, 188
Research questions, 73, 79–85, 139–148
Rhode Island, 46–47, 55–56, 60–61, 78, 89, 98, 132, 157–158, 165, 167–168, 170, 172
Royal Mathematical School (at Christ's Hospital, London), 24–30, 41, 63, 66, 70, 73, 76, 80
Royal Naval Academy, 28, 30–31, 67–68
  Portsmouth (GB), 28, 67
Rule(s) of three, 4, 8, 14–15, 17, 33, 39, 46, 49, 55–56, 59, 62, 65, 71, 78, 80–81, 91, 93, 97–98, 113, 115, 141, 151, 155–156, 187, 190
Rum, 46–47, 70
Russia, 26
Ryan, Martha and Elisabeth, 93, 148

S
Sailing, 16, 28, 63–66, 69, 160, 180
Salem, Massachusetts, 60, 82, 85, 91, 152
Schools
  abbaco, 13
  attended in winter months, 48, 106
  evening, 2–3, 42, 112, 140
  dame, 2, 17, 42, 125, 151
  for apprentices, 39, 114, 125, 140
  grade(d), 92, 125, 133, 135, 145
  grammar, 24, 42, 76, 80
  high, 2, 92, 100, 125, 135–137, 146, 149
  Latin, 20, 71, 88, 102, 120
  monitorial, 4, 121
  navigation, 31
  normal, *see* Normal schools
  private, 2, 34, 40, 42, 62, 67, 96, 121–122

Subject Index
223

public (or common), 2–3, 9, 34–35, 40, 43, 59, 62, 71, 75–76, 91, 111, 124–125, 130–133, 135, 137
reading, 73
reckoning, 9, 11, 24, 32, 37, 81, 87
rural (one-room), 107–108, 125, 145, 151
subscription, 2, 87, 91, 111, 140, 150, 166
writing, 2, 21, 29, 73, 87, 129, 148
Slates, 38, 75, 107–108, 112, 123–124, 127
Slaves, 43, 45–47, 61, 76, 95
Spain, 7, 48
Supercargo, 47
Surveying, 4–5, 12, 17, 20, 23, 25, 27, 30, 37, 40–41, 44, 60, 63, 65, 68–69, 80, 93–94, 96, 100–102, 110–111, 113, 115, 122, 141, 151, 160–161, 166, 170, 174, 178, 180
Swem Library, 72, 82, 91, 140

T

Tare and tret(t), 8, 15, 56–57, 59, 62, 70, 80–81, 98, 155, 158, 178, 189–190
Teachers
and cyphering books, 3
criticisms of, 133
in European nations, 19, 37, 40
evaluation of, 127
female and male, 20
and individual teachers, 108
navigation, 11–12, 20, 26, 30, 37, 42, 45, 67
and normal schools, 124, 131, 134–137
and penmanship, 21, 30, 77, 80, 114, 123, 127, 142, 148
qualifications of, 1–2, 34, 150
and recitation, 113–114, 116–117
salaries of, 46
and teaching methods, 1, 131, 136
work conditions of, 18
Tennessee, 89, 95, 113, 175, 182
Textbooks
on algebra, 12, 100–102
with American authors, 116
on arithmetic, 29–30, 59
authors of most widely used, 35, 68, 97, 110
and cyphering books, 107–113
with European authors, 38
expensive, 69, 114

on geometry (or Euclid), 4, 31, 69, 100–102
on navigation, 4, 100–102
scarce, 107
on surveying, 96, 100–102
on trigonometry, 4, 31, 61, 100–102
Trigonometry
oblique, 64
spherical, 63, 65
"Type" (or model) problems, 12, 15, 18, 32, 34, 73, 98, 102, 104, 111, 141, 143–144, 148

U

Understanding, 74, 106, 120, 122, 134, 146, 148–149, 151, 164
United States Hydro-graphic office, 65
University of Michigan, 31, 41, 72–74, 82, 85, 91, 107, 141
University of North Carolina (Wilson Library, Chapel Hill), 72, 82, 85, 95, 106, 140–141, 150
U.S. Department of Commerce, 136

V

Vermont Historical Society, 38
Virginia, 38, 40, 44–45, 60, 68, 72, 89, 91, 95, 99, 113, 159, 170, 180–181

W

Weights and measures, 9, 14–15, 98
West Point, US Military Academy, 31, 97, 117, 132, 135, 150
William and Mary College, 3, 45, 72, 82, 91, 106, 140, 150
Wilson Library (University of North Carolina, Chapel Hill), 72, 82, 91, 140, 150
Word problems, 12, 52, 55, 100, 166, 190
Writing-as-also-arithmetick, 29, 71, 84, 114, 139–141
Writing masters, 16, 21, 29, 31–32, 38–39, 104, 126, 140, 142, 174
Writing schools, 2, 21, 29, 73, 87, 129, 148
Written examinations, 1–2, 78, 92, 119, 127, 129, 131, 134, 145, 148
brief history of, 2

Y

Yale College (University), 45, 74, 77–78, 97, 100

Printed by Printforce, the Netherlands